# Plant Tissue Culture

# Plant Tissue Culture

**Dr. Jayarama Reddy**
Professor
St. Joseph's University
Bengaluru - 560027

## CRC Press
Taylor & Francis Group
Boca Raton London New York

CRC Press is an imprint of the
Taylor & Francis Group, an **informa** business

-EPH-
Elite Publishing House

First published 2024
by CRC Press
4 Park Square, Milton Park, Abingdon, Oxon, OX14 4RN

and by CRC Press
2385 NW Executive Center Drive, Suite 320, Boca Raton FL 33431

*CRC Press is an imprint of Informa UK Limited*

© 2024 Elite Publishing House

The right of Jayarama Reddy to be identified as author of this work has been asserted in accordance with sections 77 and 78 of the Copyright, Designs and Patents Act 1988.

Print edition not for sale in India

*British Library Cataloguing-in-Publication Data*
A catalogue record for this book is available from the British Library

ISBN13: 9781032712598 (hbk)
ISBN13: 9781032712604 (pbk)
ISBN13: 9781032712611 (ebk)

DOI: 10.4324/9781032712611

Typeset in Adobe Caslon Pro
by Elite Publishing House, Delhi

-EPH-

# Contents

| | | |
|---|---|---|
| | Preface | vii |
| | About the Author | ix |
| 1 | Introduction and History of Plant Tissue Culture | 1 |
| 2 | Terminologies Used in Plant Tissue Culture | 19 |
| 3 | Applications of Plant Tissue Culture | 25 |
| 4 | Instruments Used for Plant Tissue Culture | 33 |
| 5 | Plant Tissue Culture Laboratory Organisation | 51 |
| 6 | Basic Techniques of Plant Tissue Culture | 78 |
| 7 | Plant Tissue Culture Nutrient Media and Preparation | 96 |
| 8 | Types of Plant Tissue Culture-Organ Culture | 109 |
| 9 | Production of Haploids *in vitro* | 122 |
| 10 | Single Cell Culture | 136 |
| 11 | Cell Suspension Culture | 143 |
| 12 | Principles, Techniques of Plant Protoplast Culture | 157 |
| 13 | Agrobacterium Mediated Biotransformation | 181 |
| 14 | Bioreactors Used in Plant Tissue Culture | 196 |
| 15 | Entrepreneurship in Plant Tissue Culture | 211 |
| 16 | Automation and Robotics in Plant Tissue Culture | 222 |
| | References | 236 |

# Preface

The scenario and status Plant Tissue Culture has dramatically changed in the last two decades. It has become a multi-billion-dollar industry and a big money spinner now. There is tremendous scope and future for this technology. The entire humanity basically requires food products and medicines of good quality. The size of agricultural land almost remains the same but the population continues to grow. It will become increasingly difficult to provide the basic necessities to the ever-increasing population. Plant tissue culture will become one of the sought-after techniques to produce more food products and medicines of good quality in small spaces by making the plants more and more efficient. Plant Tissue Culture has been one of the breakthroughs in the history of plant research. It is an advanced technique of Biotechnology that allows the regeneration of a whole plant from just a few cells. It is based on a fundamental principle of totipotency. It has helped us to produce large number of plants from a small amount of source material, protect endangered species, conserve plant genomes, and proved to be an asset in horticulture, nutraceuticals, agriculture and plant-based drug manufacturing. Today, the technology is a source of earning for several giant biotech companies, low-scale companies, enthusiasts, and hobbyists. The rapid and extensive assimilation for this technology has improved the competences of the agricultural systems both in industrial and in developing countries, based on the proper application of research programs. The global plant tissue culture market size was valued at \$382.305 million in 2020, and is estimated to reach \$895.006 million by 2030, growing at a CAGR of 8.5% from 2021 to 2030. Tissue culture is the cultivation of plant cells, tissues, or organs on specially formulated nutrient media. Under the right conditions, an entire plant can be regenerated from a single cell. Plant tissue culture is a technique that has been around for more than 30 years.

This book entitled "Plant Biotechnology- Tissue Culture" is a comprehensive book on Plant Tissue Culture in the past, present & future prospects and techniques are discussed in detail. In the first three chapters, the history, terminologies and applications are given in detail. The fourth chapter is dedicated to instrumentation of plant tissue culture. The basic techniques used in PTC are described in the sixth chapter. Nutrient medium is the fulcrum of PTC. It is only by manipulating the contents of the medium, we can induce rapid and efficient growth of the plants in-vitro. The details of the constituents and types different types of nutrient media are discussed in the chapter number-8. The techniques of producing haploid plants were developed by the Indian Scientists, Guha and Maheshwari. This is one of major contributions of Indians to biology. In the chapter number 9, methods of haploid production have been described. Bioreactors are the instruments that are used for the large-scale production of plantlets and plant products. Bioreactors are developed on the fundamental principles of cell and suspension culture. These important topics are discussed in the tenth and eleventh chapters. Development of transgenic and genetically modified plants is one of major activities of biotechnology. This will yield high quality plants of rare and novel characteristics of commercial importance. Plant biotransformation and the allied techniques are discussed in the twelfth and thirteenth chapters. Medically and commercially significant plant products re produced naturally as secondary metabolites. The techniques involved their production are described the fourteenth chapter. These almost every aspect and step in Plant Tissue Culture has been automated. Robotics are being used in the large-scale production of plants and plant products. Automation and Robotics in PTC are described in the chapter number-16.

This book is useful for all the students, researchers, teachers and industrialists interested in Plant Tissue Culture. Much of the information provided in this book is based on the research and teaching experiences gained in the past three decades. The author conducted research for my Ph. D in this area at Indian Institute of Horticultural Research, Bengaluru under the able of Dr. Devinder Prakash, the then director of Orchid Laboratory at IIHR. He has doing research and teaching plant tissue culture at St. Joseph's College (Now, St. Joseph's University), Bengaluru for the last 32 years. So far, the author has published 10 books, 97 research papers and 2 patents. He has successfully guided 4 Ph. D students currently guiding another two scholars now. He collaborated with **Wageningen University, the Netherlands** for research in Gene Cloning and Bioinformatics. The author has also published 300 educational videos, which are available on his YouTube Channel-Dr. Jayarama Reddy. In 2015, National Environmental Science Academy honoured him with the **"Eminent Scientist Award-2015".** St. Joseph's College, Bangalore, presented him the **"Scroll of Honour"** for his scientific achievements in 2016. He also received G.R. Patel **"Dronacharya Award" for Academic Excellence** from International Society of Gene and Cell Therapy and Gene Research Foundation and **"J. Agarwal Gold Medal Award"** from International Society of Gene and Cell Therapy. In 2021 he received the **"Puraskara Award"** for his scientific contributions. The author has been working on the various areas of Plant Tissue Culture from the past three decades. Much of the information provided in this book is based on the research and teaching experiences gained in the past three decades. The author welcomes suggestions and comments for the improvement of the book in the next editions.

**Dr. Jayarama Reddy**

# About the Author

Dr. Jayarama Reddy is currently teaching Botany, Bioinformatics and Biostatistics for both UG and PG students at St. Joseph's University, Bengaluru-India. His field of specialization is biotechnology and he conducted research at IIHR and IISc Bengaluru for Phd. D. Dr. J. Reddy got his Ph. D (Biotechnology of Orchids) in 2002 from Bangalore University. He guided 4 Ph. D students and currently guiding another two. Dr. Jayarama Reddy has completed five research projects funded by UGC, DST etc. He has published 2 patents. Dr. Jayarama Reddy studied Magee-1, a cancer related protein and deposited it in the PDB (2H5T). He collaborated with Wageningen University, the Netherlands for research in Gene Cloning and Bioinformatics. So far, he has published 97 research papers and 9 books. In 2015, National Environmental Science Academy honoured me with the "Eminent Scientist Award-2015". St. Joseph's College, Bangalore, presented him the "Scroll of Honour" for the scientific achievements in 2016. He also received "Dronacharya Award" for Academic Excellence from International Society of Gene and Cell Therapy and Gene Research Foundation and "J. Agarwal Gold Medal Award" from International Society of Gene and Cell Therapy. In 2021 he received the "Puraskara Award". Dr. Jayarama Reddy has completed two courses in Psychology from Yale and Wesleyan Universities, USA in 2022. He has a YouTube educational channel "Dr. Jayarama Reddy". There are 300 educational videos. He is also offering online course on Narcotic Plants.

# Chapter - 1

# Introduction and History of Plant Tissue Culture

The current status of Plant Tissue Culture is quite different from the past. In the last two decades plant biotechnology applications have been widely developed and incorporated into the horticultural, industrial and medical and agricultural systems of many countries worldwide. It is fast growing to become multibillion dollar industry. Tissue culture tools have been a key factor to support such outcomes. Current results have allowed plant biotechnology and its products –including transgenic plants with several traits-to be the most assimilated technology for farmers and companies, representing several benefits. The rapid and extensive assimilation for this technology has improved the competences of the agricultural systems both in industrial and in developing countries, based on the proper application of research programs. The global plant tissue culture market size was valued at \$382.305 million in 2020, and is estimated to reach \$895.006 million by 2030, growing at a CAGR of 8.5% from 2021 to 2030. Tissue culture is the cultivation of plant cells, tissues, or organs on specially formulated nutrient media. Under the right conditions, an entire plant can be regenerated from a single cell. Plant tissue culture is a technique that has been around for more than 30 years. Tissue culture is seen as an important technology for developing countries for the production of disease-free, high quality planting material and the rapid production of many uniform plants. Micropropagation, which is a form of tissue culture, increases the amount of planting material to facilitate distribution and large-scale planting. In this way, thousands of copies of a plant can be produced in a short time. Micro-propagated plants are observed to establish more quickly, grow more vigorously and are taller, have a shorter and more uniform production cycle, and produce higher yields than conventional propagules. Plant tissue culture is a straightforward technique and many developing countries are accepting plant tissue culture for better yield. Its application only requires a sterile workplace, nursery, and green house, and trained manpower. However, tissue culture is labour intensive, time consuming, and can be costly, which will impede the market growth.

In addition, rise in developing prospects in the developed countries will further provide potential opportunities for the growth of the **plant tissue culture market** in the coming years. The rapid advancements in plant tissue culture techniques and the high demand of disease-free plants, hybrid plants and others will further accelerate the expansion of the plant tissue culture market and are also offering significant growth opportunities for the market during the forecast period.

In addition, plant tissue culture is considered to be the most efficient technology for crop improvement by the production of somaclonal and gametoclonal variants. The micropropagation technology has a vast potential to produce plants of superior quality, isolation of useful variants in well-adapted high yielding genotypes with better disease resistance and stress tolerance capacities. Hence, it will boost the market growth.

Plant tissue culture is a collection of techniques used to grow and multiply plant cells, cell aggregates, tissues or organs under sterile conditions on an artificial nutrient medium of known composition. Plant tissue culture is used to produce clones of plant in a method called micropropagation also known as *in vitro* propagation. Plant tissue culture relies on the fact that many plant cells have the ability to regenerate into a whole plant in a process called **totipotency**. Single cells, protoplasts, tissues, segments of leaves, stems or roots can often be used to generate a new plant on culture media given the required nutrients and hormones. The plant part obtained from a plant to be cultured is called explant while the main plant it is obtained from is called mother plant. Explant can be taken from different plant parts such as shoots, leaves, stems, flowers, roots, single undifferentiated cells etc. Preparation of plant tissues for tissue culture is performed under aseptic conditions under HEPA filtered air provided by laminar flow cabinet. The tissue is grown in sterile containers inside Petri dish, test tube or flasks in a growth room with controlled temperature and light intensity. Living plant materials are usually contaminated on their surfaces (or sometimes interior) with microorganisms, so their surfaces have to be sterilized using chemicals called surface sterilants. The sterile explants are placed on the solid, semi-solid or liquid media, which are generally composed of inorganic salts, organic nutrients, vitamins and plant hormones. Solid media are prepared by adding the agar the gelling agent. The composition of the medium, particularly the plant hormones and the nitrogen source have profound effects on the morphology of the tissues that grow from the initial explant. For instance, an excess of auxin will result in a proliferation of roots while an excess of cytokinin may yield to shoots proliferation.

The key factor that controls the *in vitro* morphogenetic events is the crucial ratio of auxins and cytokinins.

## Significance of Totipotency in Plant Tissue Culture (PTC)

**Totipotency is the fundamental principle of which Plant Tissue Culture.** It conveys the meaning that each and every living cell of a plant is **"Totally Potential"** enough to grow into a new plant. **Totipotency** is the genetic potential of a plant cell to produce the entire plant. The basis of tissue culture is to grow large number of cells in a sterile controlled environment. The cells are obtained from stem, root or other plant parts and are allowed to grow in culture medium containing mineral nutrients, vitamins and hormones to encourage cell division and growth. As a result, the cells in culture will produce an unorganised proliferative mass of cells which is known as callus tissue. The cells that comprise the callus mass are totipotent. Thus, a callus tissue may be in a broader sense totipotent, i.e., it may be able to regenerated back to normal plant given certain manipulations of the medium and the cultural environment. Truly speaking, totipotency of the cell is manifested through the process of differentiation and the hormones in this process play the major role than any other manipulations. The first attempt of plant tissue culture was made in 1902 by a German physiologist, Gottlieb Haberlandt. He tried to culture isolated single palisade cells from leaves in knop's salt solution enriched with sucrose but failed. He then, envisaged the need of stimulants (later known as hormones) that were lacking in the medium used by him. In the fifties, Skoog and Miller discovered a new plant growth hormone kinetin from herring sperm DNA. With a correct concentration ratio of auxin and cytokinin in tobacco cultures, Skoog was able to demonstrate the role of kinetin in organogenesis. When the ratio of kinetin to auxin was higher, only shoot developed. This is known as caulogenesis. But when the ratio was lower, only roots were formed. This is known as rhizogenesis. Around the same period, Steward et al devised a method for growing carrot tissue by excising small disc, from the secondary phloem region of carrot root and placing them in a moving liquid medium under aseptic conditions. In presence of coconut milk in the medium, the phloem tissue began to the grow actively.

In moving liquid medium some single cells and small groups of cells were loosened from the surface of growing tissue. When these isolated cells were grown separately it was found that some single cells developed somatic embryos or embryoids by a process that occurs in normal zygotic embryo. It is also observed in some experiment that cells of some callus mass frequently differentiate into vascular elements such as xylem and phloem without forming any plant organs or embryoids. This

4

process is known as histogenesis or Cyto-differentiation. Thus, the totipotent cells may express themselves in different way on the basis of differentiation process and manipulation. Where the totipotent cells are partially expressed or not expressed, it is obvious that the limitation on its capacity for development must be imposed by the microenvironments. The totipotency of cells in the callus tissue may be retained for a longer period through several subcultures. Practically, it is observed that the ex- plant first forms the callus tissue in the callus inducing medium and such callus tissue is maintained through some subcultures. After then it is generally transferred to another medium which is expected to be favourable for the expression of totipotent cells. Actually, the regeneration medium is standardized by trial-and-error method. In more or less suitable medium, the totipotent cells of the callus tissue give rise to meristematic nodules or meristemoids by repeated cell division. This may subsequently give rise to vascular differentiation or it may form a primordium capable of giving rise to a shoot or root. Sometimes the totipotent cell may produce embryoids through sequential stages of development such as globular stage, heart shaped stage and torpedo stage etc. After prolonged culture, it has been observed that callus in some species (e.g., *Ntcotiana tabacum, Citrus aurantifolia* etc.) maybe- come habituated. This means that they are now able to grow on a standard maintenance medium which is devoid of growth hormones. The cells of habituated callus also remain totipotent and are capable to regenerate a plant without any major manipulation.

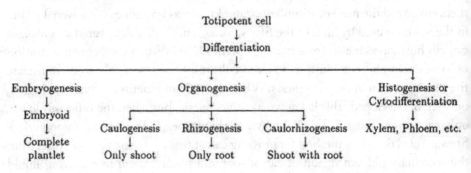

A typical crown gall tumour cell has the capacity for unlimited growth independent of exogenous hormones. It shows totally lack of organ genic differentiation. So, such tissue is considered to have permanently lost the totipotentiality of the parent cells. In some plant species, the crown gall bacterium (*Agrobactenum tumefaciens*) induces a special type of tumour, called teratomas, the cells of which possess the capacity to differentiate shoot buds and leaves when they are grown in culture for unlimited periods. Thus, it is clear that the mode of expression of totipotency of plant cell in culture varies from plant to plant and also helps us to understand the process of differentiation in vitro.

## Importance of Totipotency in PTC

The ultimate objective in plant protoplast, cell and tissue culture is the reconstruction of plants from the totipotent cell. Although the process of differentiation is still mysterious in general, the expression of totipotent cell in culture has provided a lot of information. On the other hand, the totipotentiality of somatic cells has been exploited in vegetative propagation of many economical, medicinal as well as agriculturally important plant species. Therefore, from fundamental to applied aspect of plant biology, cellular totipotency is highly important. Recent trends of plant tissue culture include genetic modification of plants, production of homozygous diploid plants through haploid cell culture, somatic hybridization, mutation etc. The success of all these studies depends upon the expression of totipotency. In many cases, successful and exciting results have been obtained. Plant breeders, horticulturists and commercial plant growers are now more interested in plant tissue culture only for the exploitation of totipotent cells in culture according to their desirable requirement. Totipotent cells within a bit of callus tissue can be stored in liquid nitrogen for a long period. Therefore, for germplasm preservation of endangered plant species, totipotency can be utilized successfully.

PTC can be defined as the culture of all types of plant cells, tissues and organs under aseptic, controlled nutritional and environmental conditions often to produce the clones of plants. PTC also encompasses the culture of excised other plant organs such as floral parts, anthers, embryos and protoplast culture. Plant tissue culture has an important role to play in the manipulation of plants for improved agronomic performance. Plant tissue culture is an integral part of molecular approaches to plant improvement and acts as an intermediary whereby advances made by the molecular biologists in gene isolation and modification are transferred to plant cells. Most applied and well translated among these is the technique of micropropagation, which has revolutionized the modern horticulture and agriculture industries. Based on the availability of the various *in vitro* techniques, the dramatic increase in their application to various problems in basic biology, agriculture, horticulture, and forestry. The applications can divide conveniently into seven broad areas;

1. Cell behavior and plant nutritional studies

2. Genetic modification of plants

3. Germplasm storage and cryopreservation

4. Clonal propagation and mass multiplication.

5. Secondary metabolites

6. Production of Improved verities.

7. Production of pathogen free plants.

The resultant clones of plants produced through PTC are true-to type of the selected genotype. The controlled conditions provide the culture an environment conducive for their growth and multiplication. These conditions include proper supply of nutrients, pH medium, adequate temperature and proper gaseous and liquid environment. PTC is being widely used for large scale plant multiplication. Apart from their use as a tool of research, plant tissue culture techniques have in recent years, become of major industrial importance in the area of plant propagation, disease elimination, plant improvement and production of secondary metabolites. Small pieces of tissue (named explants) can be used to produce hundreds and thousands of plants in a continuous process. A single explant can be multiplied into several thousand plants in relatively short time period and space under controlled conditions, irrespective of the season and weather on a year-round basis. Endangered, threatened and rare species have successfully been grown and conserved by micropropagation because of high coefficient of multiplication and small demands on number of initial plants and space.

In addition, plant tissue culture is considered to be the most efficient technology for crop improvement by the production of somaclonal and gametoclonal variants. The micropropagation technology has a vast potential to produce plants of superior quality, isolation of useful variants in well-adapted high yielding genotypes with better disease resistance and stress tolerance capacities. Certain type of callus cultures give rise to clones that have inheritable characteristics different from those of parent plants due to the possibility of occurrence of somaclonal variability, which leads to the development of commercially important improved varieties. Commercial production of plants through micropropagation techniques has several advantages over the traditional methods of propagation through seed, cutting, grafting and air-layering etc. It is rapid propagation processes that can lead to the production of plants virus free. *Coryodalis yanhusuo*, an important medicinal plant was propagated by somatic embryogenesis from tuber-derived callus to produce disease free tubers. Meristem tip culture of banana plants devoid from banana bunchy top virus (BBTV) and brome mosaic virus (BMV) were produced. Higher yields have been obtained by culturing pathogen free germplasm *in vitro*. Increase in yield up to 150% of virus-free potatoes was obtained in controlled conditions. The main objective of writing this chapter is to describe the tissue culture techniques, various developments, present and future trends and its application in various fields.

## Advantages of PTC

1. Rapid and large-scale production of commercially important mature plants.

2. The production of multiples of plants in the absence of seeds or pollinators to produce seeds.

3. The production of exact copies of plants that produce particularly good flowers, fruits or have other desirable traits.

4. The production of plants in sterile conditions with greatly reduced chances of transmitting diseases, pests and pathogens.

5. The production of plants from seeds that otherwise have low chances of germinating and growing.

6. To mass propagate plants for commercial use and biosynthesis of secondary metabolites using bioreactors.

7. It also produces disease-free plants due to its method of growth.

## Limitations of PTC

1. The setting up of a plant tissue culture laboratory is very expensive including the machines and reagents.

2. The experiments of tissue culture must be handled by highly trained people as the procedure requires special care and careful observations.

3. If all the plants are genetically similar, there is reduction in genetic diversity.

4. If a plant is susceptible to disease, all the plants of this cloned stock will share this undesirable trait and be susceptible to that particular disease.

5. The procedures depends on the type of species being cultured, hence there is need for trial-and-error method for any new species if there is no review about that species.

6. If precautions are not taken, the whole stock may be contaminated or infected.

7. High venture capital and big risk are involved in setting up a commercial plant tissue culture industry. Marketing the PTC grown plants and products is a tricky process and needs technical expertise.

## Major Steps in PTC

There are four major *in vitro* steps in PTC for the purpose of micropropagation or mass multiplication of plants. But for the purpose of the synthesis of secondary metabolites there are only two steps. A preliminary step is most important before one begins a project on PTC weather it is micropropagation or secondary metabolite synthesis, **this step is compulsory.**

**Preliminary step:** It is a very crucial step in which all the necessary decisions are taken. First of all, the plant to be cultured is selected. The selection of the plant species is done based on importance, need, economic viability and feasibility. Later, based on the nature of the plant whether it is an herb or shrub or tree the selection of the nutrient media is made. There thousands of nutrient media. Even the selection of the explant to be used in culture is also crucial. For taking all these crucial decisions one should have a proper knowledge about PTC technology. Besides this a thorough literature and economic survey is essential.

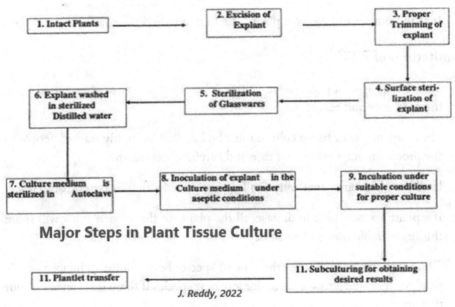

**Major Steps in Plant Tissue Culture**

*J. Reddy, 2022*

### Steps in Plant Tissue Culture

**Steps in micropropagation or clonal propagation or mass multiplication:**

**Stage I**-A step in *in vitro* propagation characterized by the establishment of an aseptic tissue culture of a plant. This involves media preparation and inoculation.

The main purpose here is to raise healthy plantlets in large numbers, which can be achieved either by **direct organogenesis or indirect organogenesis**.

**Stage II**-A step in *in vitro* propagation characterized by the rapid numerical increase of organs or other structures. This is carried out by **subculturing**.

**Stage III**- This step is called acclimatization and this *in vitro* propagation step is characterized by preparation of propagules for successful transfer to soil, a process involving rooting of shoot cuttings, hardening of plants, and initiating the change from the heterotrophic to the autotrophic state.

**Stage IV**-A step in in vitro plant propagation characterized by the establishment in soil of a tissue culture derived plant, either after undergoing a Stage III pre-transplant treatment, or in certain species, after the direct transfer of plants from Stage II into soil.

**Steps in secondary metabolite synthesis:**

**Step I**- A step in *in vitro* propagation characterized by the establishment of an aseptic tissue culture of a medicinally or commercially important plants (e.g., *Catharanthus, Ginseng, Taxus* etc.). This involves media preparation and inoculation. This is same as in the case of micropropagation but the purpose here is to raise callus and not plantlets.

**Step II**- Commercial multiplication and maintenance of the callus. This is generally done in large and industrial scale in bioreactors and continuous cultures. Extraction of medicinally or commercially important plant products or secondary metabolites.

**Brief Historical Account**

Plant cell, tissue and organ culture, also referred to as *in vitro*, axenic, or sterile culture, has now become an important tool in both basic and applied studies as well as in commercial application. The earliest step toward plant tissue culture was made by Henri-Louis Duhumel du Monceau in 1756, who, during his pioneering studies on wound-healing in plants, observed callus formation. The science of plant tissue culture takes its roots from the discovery of cell followed by propounding of cell theory. In 1838, Schleiden and Schwann proposed that cell is the basic structural unit of all living organisms. They visualized that cell is capable of autonomy and therefore it should be possible for each cell if given an environment to regenerate into whole plant. Based on this premise, in 1902 Gottlieb Haberlandt for the first time attempted to culture isolated single palisade cells from leaves in knop's salt solution enriched with sucrose. The cells remained alive for up to one month, increased in size, accumulated starch but failed to divide. Though he was unsuccessful but laid down the foundation of tissue culture technology for which he is regarded as the father of plant tissue culture. He opined that to "my knowledge, no systematically organized attempts to culture isolated vegetative cells from higher plants have been made. Yet the results of such culture experiments should give some interesting insight to the properties and potentialities which the cell as an elementary organism possesses. Moreover, it would provide information about the inter-relationships and complementary influences to which cells within a multicellular whole organism are exposed".

Embryo culture had its beginning early in the nineteenth century, when Hannig in 1904 successfully cultured cruciferous embryos. Using a different approach Kotte (1922), a student of Haberlandt, and Robbins (1922) succeeded in culturing isolated root tips. This approach, of using explants with meristematic cells, led to the successful and indefinite culture of tomato root tips by White in 1934. Further studies allowed for root culture on a completely defined medium. The first true plant tissue cultures were obtained by Gautheret in 1935 from cambial tissue of *Acer pseudoplatanus*. He also obtained success with similar explants of *Ulmus campestre, Robinia pseudoacacia,* and *Salix capraea* using agar-solidified medium of Knop's solution, glucose, and cysteine hydrochloride.

The 1940s, 1950s, and 1960s proved an exciting time for the development of new techniques and the improvement of those already available. The application of coconut water by Van Overbeek et al. in 1941 allowed for the culture of young embryos and other recalcitrant tissues, including monocots. Discovery of kinetin by Miller in 1955 further increased the number of species that could be cultured indefinitely. Subsequently the discovery of other hormones and other complex adjuvants led to the solid establishment of PTC technique.

Murashige and Skoog in 1962 developed a new medium. The concentration of some salts were 25 times that of Knop's solution. In particular, the level of $NO_3^-$ and $NH_4^+$ were very high and the array of micronutrients were increased. This formulation allowed for a further increase in the number of plant species that could be cultured, many of them using only a defined medium consisting of macro- and micronutrients, a carbon source, reduced nitrogen, B vitamins, and growth regulators. This medium is very popular in PTC and is extensively used even in the PTC industry. The culture of single cells which is now a very popular industrial technique for the production of plant-based drugs other secondary metabolites was achieved by Muir et al in 1954 through shaking callus cultures of *Tagetes erecta* and tobacco and subsequently placing them on filter paper resting on well-established callus, giving rise to the so-called nurse culture. Central to the success of producing biologically active substances commercially is the capacity to grow cells on a large scale. This is now being achieved using bioreactors, stirred tank reactor systems and a range of air-driven reactors. Techniques for *in vitro* culture of floral and seed parts were developed in 1942 by LaRue. *In vitro* pollination and fertilization were pioneered by Kanta et al.in 1962 using *Papaver somniferum*. The preliminary achievements with respect to the protoplast culture were made by Cocking in 1960. *In vitro* techniques now have truly revolutionized agriculture due because of the initial success achieved by Morel in 1960s.

Using tissue culture for conservation began in the mid-1980s when assistance was provided by international agencies for collecting root crops. By 1987 collections existed in many countries, and in several cases, efforts had been made to characterize and evaluate the collections. This germplasm had to be shared, and safely, in a region composed of island countries, each with strict quarantine regulations. Using tissue culture enabled a number of requirements to be met. Through meristem culture, plants free of viruses could be produced, which facilitates distribution, as quarantine concerns are addressed. Tissue culture can be used to improve on conventional multiplication rates, which satisfies the need to provide a sufficient volume of material to each country. Finally, the problems that can be incurred when sending conventional vegetative propagules do not arise when distributing tissue cultures. Tissue culture laboratories were therefore established within regional institutes, and pathogen-testing schemes developed. A tissue culture laboratory was established at the then, South Pacific Commission (SPC), in Suva, Fiji, maintaining pathogen-tested collections of taro, cassava, yam, banana, sweet potato and vanilla. In 1986 the Commonwealth Fund for Technical Cooperation (CFTC) established a tissue culture laboratory at the University of the South Pacific (USP), Samoa. In 1990, the European Union, through the Pacific Regional Agricultural Programme (PRAP), started funding this laboratory, as part of the PRAP project, 'Provision of Tissue Culture Services for the Region'. The laboratory was modified to conserve some 20,000 cultures under normal and slow growth conditions. The EU funded this project until December 1999. The PRAP laboratory worked closely with the SPC laboratory in germplasm distributions and research into conservation and multiplication methodologies.

During the 1990s and the early twenty-first century continued expansion in the application of *in* vitro technologies to an increasing number of plant species has been observed. Tissue culture techniques are being used with all types of plants, including cereals and grasses (Vasil, 1994). Legumes (Davey et al., 1994), vegetable crops (Reynolds, 1994), potato (Jones, 1994) and other root and tuber crops (Krikorian, 1994a), oilseeds (Palmer & Keller, 1994), temperate (Zimmerman & Swartz, 1994) and tropical (Grosser, 1994) fruits, plantation crops (Krikorian, 1994b), forest trees (Harry & Thorpe, 1994), and ornamentals (Debergh, 1994).

An important output of both laboratories was in raising the awareness of the benefits tissue culture can bring to agriculture. These benefits range from improved propagation rates to distribution of pathogen-tested germplasm to safe and secure conservation. Research conducted at the PRAP laboratory led to the development of a micropropagation technique, which significantly improved on the multiplication rates obtained with taro conventionally. This method is now being used by the Ministry of Agriculture in their national tissue culture laboratory in Samoa for bulking up improved, taro leaf blight resistant varieties, so they can be made available to the farmers. The role that tissues culture can play in a conservation strategy was demonstrated when looking at the losses that had occurred throughout the region in field gene-banks. Many of the accessions lost in the field gene-banks, were safely conserved in tissue culture. Finally, access to pathogen-tested germplasm was shown to be very important in a region, composed of many islands, each with their own strict quarantine regulations. The existence of a regional tissue culture laboratory, maintaining pathogen-tested tissue cultured germplasm therefore facilitated distribution and utilization, as well as being the best option for conservation.

The need for a regional tissue culture laboratory was reinforced with the initiation of the Taro-Gen project. This project was establishing national field collections of taros throughout the region, from which a core collection, representative of the genetic diversity of the whole collection, would be established. This core collection had to be safely conserved for utilization by taro growers in the region. Safe and secure conservation could only be guaranteed by using a regional tissue culture facility. In addition, no IARC has the mandate to conserve taro, and so it was felt that there should be a facility responsible for the conservation, improvement and utilization of this very important Pacific crop.

Advances in molecular biology allow for the genetic engineering of plants through the precise insertion of foreign genes from diverse biological systems. Three major breakthroughs have played major roles in the development of this transformation technology. These are the development of shuttle vectors for harnessing the natural gene transfer capability of *Agrobacterium*, the methods to use these vectors for the direct transformation of regenerable explants obtained from plant organs, and the development of selectable markers. For species not amenable to Agrobacterium-mediated transformation, physical, chemical, and mechanical means are used to get the DNA into the cells. With these latter approaches, particularly biolistics, it is becoming possible to transform any plant species and genotype. The initial wave of research in plant biotechnology has been driven mainly by the seed and agrichemical industries and has concentrated on "agronomic traits" of direct relevance to these industries, namely the control of insects, weeds, and plant diseases. At present, several hundred crop species of plants have been genetically engineered, including nearly all the major dicotyledonous families. An increasing number of monocotyledonous ones as well as some woody plants. Current research has led to routine gene transfer systems for most important crops. In addition, technical improvements are further increasing transformation efficiency, extending transformation to elite commercial germplasm and lowering transgenic plant production costs. The next wave in agricultural biotechnology is already in progress with biotechnological applications of interest to the food processing, speciality chemical, and pharmaceutical industries.

**Outstanding Contributors to PTC:** The following scientists have laid the solid foundation to the field of PTC.

## Wilhelm Knop (1817-1891)

During 1859-1865, Knop prepared a nutrient solution used in the growth of higher plants in water, the technique called hydroponics. The solution contained definite proportions of calcium nitrate, potassium nitrate, magnesium sulfate, monobasic potassium phosphate, and potassium chloride dissolved in water. This solution is known to be the basis of all the nutrient media used in PTC today.

**Gottlieb Haberlandt** (1854- 1945), an Austrian botanist is known as the father of plant tissue culture. He was the son of European 'soybean' pioneer Professor Friedrich J. Haberlandt. His son Ludwig Haberlandt was an early reproductive physiologist now given credit as the 'grandfather' of the birth control pill, the pill. Haberlandt first pointed out the possibilities of the culture of isolated tissues, plant tissue culture. He suggested that the potentialities of individual cells via tissue culture and also suggested that the reciprocal influences of tissues on one another could be determined by this method. Since Haberlandt's original assertions methods for tissue and cell culture have been realized, leading to significant discoveries in Biology and Medicine. His original idea presented in 1902 was called totipotentiality: "Theoretically all plant cells are able to give rise to a complete plant." The more efficient C-4 photosynthesis in land plants depends on a specialized Kranz (German for wreath) leaf anatomy History of C3: C4 photosynthesis research first described by Gottlieb Haberlandt in 1904.

Haberlandt decided that his students would profit from a system of classifying plants based on function. In his book *Physiologische Pflanzenanatomie* (1884; "Physiological Plant Anatomy") he distinguished 12 tissue systems based on function (mechanical, absorptive, photosynthetic, etc.). Although his system was not accepted by other botanists, the analysis of the relations between structure, function, and environment has been useful in the study of plant adaptations to different habitats.

**P R White** is known for the first continuous culture of plant tissues in vitro, which he achieved in 1934. He demonstrated the first indefinite culture of tomato roots. He is also known for nutrient medium, which is popularly known as White's medium.

Greats of PTC: 1-Wilhelm Knop; 2-Haberlandt; 3-PR White; 3-Georges Morel; 4-Georges Morel; 5-Folke Skoog; 6-Toshio Murashige

**Important Scientists of Plant Tissue Culture**

**George Michel Morel** (1916-1973), a plant physiologist at the National Institute for Agronomic Research in France, was one of many scientists who had become interested in the formation of tumors in plants as well as in studying various pathogens such as fungi and viruses that cause plant disease. He was successful in culturing tissue from ferns and was the first to culture monocot plants. He utilized a small piece of the growing tip of a plant shoot (the shoot apex) as the starting tissue material. This tissue was placed in a glass tube, supplied with a medium containing specific nutrients, vitamins, and plant hormones, and allowed to grow in the light. Under these conditions, the apex tissue grew roots and buds and eventually developed into a complete plant. He was able to generate whole plants from pieces of the shoot apex that were only 100 to 250 micrometers in length. He also investigated the growth of parasites such as fungi and viruses in dual culture with host-plant tissue. Using results from these studies and culture techniques that he had mastered, Morel and his colleague Claude Martin regenerated virus-free plants from tissue that had been taken from virally infected plants. He was the first to recognize the potential of the in vitro culture methods for the mass propagation of plants. He estimated that several million plants could be obtained in one year from a single small piece of shoot-apex tissue. Plants generated in this manner were clonal (genetically identical organisms prepared from a single plant.

**Folke Karl Skoog and Toshio Murashige** *formulated the famous*, Murashige and Skoog medium (or MSO or MS0 (MS-zero). It is a plant growth medium used extensively in the laboratories for cultivation of plant cell culture. MSO was invented by plant scientists Toshio Murashige and Folke K. Skoog in 1962 during Murashige's search for a new plant growth regulator. A number behind the letters MS is used to indicate the sucrose concentration of the medium. For example, MS0 contains no sucrose and MS20 contains 20 g/l sucrose. Along with its modifications, it is the most commonly used medium in plant tissue culture experiments in laboratory. As Skoog's doctoral student, Murashige originally set out to find an as-yet undiscovered growth hormone present in tobacco juice. No such component was discovered; instead, analysis of juiced tobacco and ashed tobacco revealed higher concentrations of specific minerals in plant tissues than were previously known. A series of experiments demonstrated that varying the levels of these nutrients enhanced growth substantially over existing formulations. It was determined that nitrogen in particular enhanced growth of tobacco in tissue culture.

## Major Mile-stones in PTC

1892- Klercker: Plant protoplasts or cells without cell walls were first mechanically isolated from plasmolyzed tissues.

1904-Hanning: Embryo culture also had its beginning early in the nineteenth century, when Hannig in 1904 successfully cultured cruciferous embryos and Brown in 1906 barley embryos

1909- Küster: Achieved the first fusion of isolated protoplast.

1922-Kotte: Using a different approach a student of Haberlandt, and Robbins (1922) succeeded in culturing isolated root tips.

1934-35-Gautheret: The application of coconut water (often incorrectly stated as coconut milk) was envisaged.

1934-Tukey: was able to allow for full embryo development in some early-ripening species of fruit trees, thus providing one of the earliest applications of in vitro culture

1941-Braun: showed that *Agrobacterium tumefaciens* could induce tumors in sunflower, not only at the inoculated sites, but at distant points. These secondary tumors were free of bacteria and their cells could be cultured without auxin.

1941-Van Overbeek: Coconut water was used in the culture of young embryos and other recalcitrant tissues, including monocots.

1946-Ball: successfully produced plantlets by culturing shoot tips with a couple of primordia of Lupinus and Tropaeolum.

1954-1958- Muir et al : The culture of single cells (and small cell clumps) was achieved by shaking callus cultures of *Tagetes erecta* and tobacco and subsequently placing them on filter paper resting on well-established callus, giving rise to the so-called nurse culture.

1954-Skoog: showed that the addition of adenine and high levels of phosphate allowed nonmeristematic pith tissues to be cultured and to produce shoots and roots.

1957-Skoog & Miller- The availability of kinetin further increased the number of species that could be cultured indefinitely, but perhaps most importantly, led to the recognition that the exogenous balance of auxin and kinetin in the medium influenced the morphogenic fate of tobacco callus.

1959-Kohlenbach: finally succeeded in the culture of mechanically isolated mature differentiated mesophyll cells of Macleaya cordata and later induced somatic embryos from callus.

1960: Protoplast isolation and culture remained an unexplored technology until the use of a fungal cellulase by Cocking (1960) ushered in a new era.

1960-Morel: The importance of PTC to obtain virus-free orchids, realized its potential for clonal propagation. It triggered a sort of revolution in PTC resulting popularization of the technique.

1962- Kanta et al: *In vitro* pollination and fertilization was pioneered by Kanta et al. (1962) using *Papaver somniferum*.

1962-Murashige and Skoog (1962): developed a new medium, which is now being widely used for the propagation of most of the herbaceous plants. The concentration of some salts were 25 times that of Knop's solution. In particular, the level of $NO^3$- and $NH^{4+}$ were very high and the array of micronutrients were increased.

1964-66-Guha and Maheshwari: Produced haploid plants from cultured anthers of *Datura innoxia* that opened the new area of androgenesis.

1970s: The commercial availability of cell-wall-degrading enzymes led to their wide use and the development of protoplast technology in the 1970s.

1972- Carlson et al: Achieved the regeneration fused protoplasts of the first interspecific hybrid plants (*Nicotiana glauca* × *Nicotiana langsdorffii*).

1980-200: Dramatic increase in the use of PTC in 1980-2000 and the techniques has been used extensively in five broad areas, namely: (a) cell behavior, (b) plant modification and improvement, (c) pathogen-free plants and germplasm storage, (d) clonal propagation, and (e) product formation.

1980s: Genetic modification of plants is being achieved by direct DNA transfer via vector-independent and vector-dependent means since the early 1980s.

1990s: The use of *Agrobacterium* in vector-mediated transfer has progressed very rapidly since the first reports of stable transformation in 1990s.

2000 onwards: cell suspension cultures, shoot apices, asexual embryos, and young plantlets, after treatment with a cryoprotectant, are frozen and stored at the temperature of liquid nitrogen (ca. −196°C). It gave way for cryopreservation and conservation strategies.

In the twenty-first century continued expansion in the application of *in vitro* technologies to an increasing number of plant species has been observed. Tissue culture techniques are being used with all types of plants, including cereals and grasses, legumes, vegetable crops, potatoes and other root and tuber crops, oilseeds, fruits, plantation crops, forest trees, ornamentals and large-scale industrial production of

18

secondary metabolites of industrial and pharmaceutical importance. The concurrent progress in applied plant biotechnology is fully matching and is in fact stimulating fundamental scientific progress. As a result of which PTC racing towards becoming Billion Dollar Business.

# Chapter - 2

# Terminologies Used
# in Plant Tissue Culture

**Adventitious structures:** Developing from unusual points of origin, such as shoot or root tissues, from callus or embryos, from sources other than zygotes.

**Agar:** It is a gelling agent of algal in origin. It is a polysaccharide powder used for making the solid medium at a concentration of 6-12 g/litre.

**Artificial Nutrient Medium:** It is a nutritive solution for culturing cells in which each component is specifiable and ideally of known chemical structure. E.g., Murashige and Skoog Medium.

**Aseptic conditions:** Procedures used to prevent the introduction of fungi, bacteria, viruses, mycoplasma or other microorganisms into cultures.

**Autoclave:** A machine capable of sterilizing wet or dry items with steam under pressure. Pressure cookers are a type of autoclaves.

**Callus:** It is an unorganized, proliferate mass of differentiated plant cells, a wound response.

**Caulogenesis:** Type of organogenesis by which only adventitious shoot bud initiation take place in the callus tissue.

**Clonal Propagation:** Asexual reproduction of plants that are considered to be genetically uniform and originated from a single individual or explant. Nowadays it is being widely used in the multiplication of crop plants.

**Competency** is the endogenous potential of a given cell or tissue to develop into a complete plant under *in vitro* conditions.

**Contamination:** Accidental growth of unwanted microorganisms such as bacteria or fungi in the culture vessels.

**Cytodifferentiation**: In plant tissue culture, during growth and maturation of the callus tissue or free cells in suspension culture, few dedifferentiated cells undergo cyto-quiescence and cyto-senescence and these twin phenomena are mainly associated with differentiation of vascular tissue, particularly tracheary elements. The whole developmental process is termed as Cytodifferentiation. The fate of an individual cell in culture is variable and hence unpredictable. Amongst a group of cells within the callus tissue or free cells in cell suspension culture, a few cells become morpho-genetically competent for Cytodifferentiation which cannot be identified at the early stage in advance. Cytodifferentiation occurs either spontaneously or under the stimulus of specific nutritional or hormonal factors. So, it is not conditioned by a single regular event.

**Determinism** *is the quality of cells or tissues* in the process of in vitro shoot organogenesis.

**Differentiation Dedifferentiation and Redifferentiation:** The cells derived from root apical and shoot-apical meristems and cambium differentiate and mature to perform specific functions. This act leading to maturation is termed as differentiation. During differentiation, cells undergo few to major structural changes both in their cell walls and protoplasm. For example, to form a tracheary element, the cells would lose their protoplasm. They also develop a very strong, elastic, lignocellulosic secondary cell walls, to carry water to long distances even under extreme tension. Try to correlate the various anatomical features you encounter in plants to the functions they perform. Plants show another interesting phenomenon. The living differentiated cells that by now have lost the capacity to divide can regain the capacity of division under certain conditions. This phenomenon is termed as dedifferentiation. For example, formation of meristems – interfascicular cambium and cork cambium from fully differentiated parenchyma cells. While doing so, such meristems / tissues are able to divide and produce cells that once again lose the capacity to divide but mature to perform specific functions, i.e., get redifferentiated. List some of the tissues in a woody dicotyledenous plant that are the products of redifferentiation.

**Embryoids**: Somatic embryogenesis is an artificial process in which a *plant* or embryo is derived from a single somatic cell or group of somatic cells. Somatic embryos are formed from *plant* cells that are not normally involved in the development of embryos. The embryo like cell aggregates thus raised in vitro are called embryoids.

**Haploid plants:** Development of haploid plants from immature pollen grains of an anther of *Nicotinia tabaccum.* was first discovered in 1966 by Guha and Maheshwari when studying meiosis *in vitro* in *Datura innoxia* anthers. Sometimes differentiation takes place in the absence of cell division: in tissue culture systems, such as *Zinnia elegans*, we commonly find xylem elements appearing among otherwise undifferentiated cells. This form of xylogenesis represents the acquisition of a specific metabolic competence that is quite different from that of the parental cell. Clearly, differentiation *in vitro* can take several forms.

**In vitro:** Meaning–in glass (Latin). It refers to the propagation of plants in a controlled, artificial environment using plastic or glass culture vessels, aseptic techniques, and a defined growing medium.

**Laminar Air Flow (LAF):** An enclosed work area that has sterile air moving across it. The air moves with uniform velocity along parallel flow lines. Room air is pulled into the unit and forced through a HEPA (High Energy Particulate Air) filter, which removes particles 0.3 μm and larger.

**Meristemoids:** It is localized group of meristematic cells that arise in callus tissue and may give rise to shoots and or roots. They are also termed as nodules or growth centers.

**Micropropagation:** *In vitro* clonal propagation of plants from shoot tips or nodal explants, usually with an accelerated proliferation of shoots during subcultures.

**Organogenesis:** is the development of adventitious organs or primordia from undifferentiated cell mass in tissue culture by the process of differentiation is called organogenesis. The formation of roots, shoots or flower buds from the cells in culture in manner similar to adventitious root or shoot formation in cuttings is called organogenesis.

**Organoids:** In some culture tissues, an error occurs in development programming for organogenesis and an anomalous structure is formed. Such anomalous organs like structures are known as organoids. Although or ganoids contain the dermal, vascular and ground tissues present in plant organs, they differ from true organ in that the organoids are formed directly from the periphery of the callus tissue and not from organized meristemoids.

**Plant Growth Regulators (Hormones):** Chemicals that are included in the nutrient media. They may synthetic or natural in origin and strongly affects growth (cytokinin, auxins, and gibberellins).

**Protocorm Like Bodies (PLBs):** When an orchid seed germinates in nature generally a Protocorm is produced in natural conditions. Similarly, if the orchid seeds are germinated *in vitro*, they are called PLBs. They can be also be induced to grow from an orchid explant under *in vitro* conditions.

**Rhizogenesis:** Type of organogenesis by which only adventitious root formation takes place in the callus tissues.

**Somaclonal variations:** Phenotypic variation, either genetic or epigenetic in origin, displayed among somaclones or plants derived from any form of cell culture involving the use of somatic plant cells.

**Totipotency:** It is the genetic potential of a plant cell to produce the entire plant. In other words, totipotency is the cell characteristic in which the potential for forming all the cell types in the adult organism is retained.

**Xylogenesis:** The process of development and differentiation of xylem elements is known as

**Types of differentiation:** Differentiation implies development of organised structures, usually from undifferentiated tissue but also from previously specialised cells that would not normally give rise to organised multicellular growth (epidermal cells, pollen grains). In plant tissue culture, undifferentiated tissue is referred to as callus (Figure 1a) although a callus can contain meristematic nodules that may not be obvious to the naked eye but which never develop further unless suitable conditions are supplied. Development of organised structures can follow one of three pathways **(See Fig-below)**

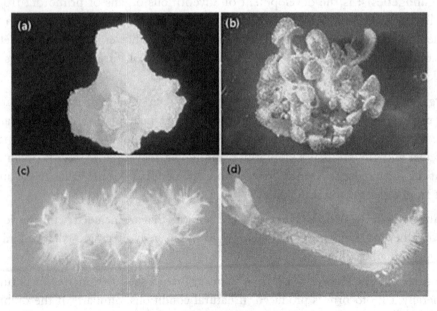

**In Vitro Morphogenetic stages**

## *In Vitro* Morphogenesis (See Fig Above)

1. Undifferentiated tissue or callus formation from the explant (a)

2. Shoot regeneration, based on a unipolar structure with a shoot apical meristem (b)

3. Root regeneration, essentially a unipolar structure with a root apical meristem (c)

4. Somatic embryogenesis in which there is a bipolar structure (d).

**Indirect organogenesis:** plant growth regulators and differentiation: An understanding of the mechanisms underlying regeneration of whole plants, or parts of plants, from cells has come some way since the classic observations of Skoog and Miller that the direction of differentiation could be influenced by the ratio of the exogenously supplied growth regulators auxin and cytokinin. They observed in tobacco stem pith cultures that a high ratio of auxin to cytokinin led to initiation of roots whereas a low ratio led to development of shoots. Although there are many species for which this simple manipulation will not work, in general auxins (e.g. IAA, indoleacetic acid), NAA (a-naphthaleneacetic acid) and IBA indole butyric acid)) will stimulate regeneration of roots, and cytokinins (e.g. BAP (6-benzylaminopurine) and kinetin) will promote regeneration of shoots or embryos. Although auxin stimulates initial cell division in quiescent cells, continued presence of auxin can inhibit organised out-growth. This is a typical example of the sequential functions of a single hormone through a developmental progression. In practical terms, cultures are usually transferred onto low or zero auxin media to permit or speed up shoot organogenesis. Sometimes 'removal' of auxin occurs when auxin in the medium is degraded either by the tissue itself or via chemical reactions such as photo-oxidation. Cytokinins promote out-growth of shoots but are normally kept at very low concentration when root regeneration is wanted.

**Direct organogenesis:** the role of growth regulators: Direct organogenesis bypasses the need for a callus phase. A good example is the formation of somatic embryos. Most evidence suggests that direct embryogenesis proceeds from cells which were already embryogenically competent while they were part of the original, differentiated tissue. These pre-embryogenic cells appear only to require favourable conditions (such as wounding or application of exogenous growth regulators) to allow release into cell division and expression of embryogenesis. Such cells tend to be much more responsive than those involved in indirect organogenesis and do not seem to require the same auxin 'push' to initiate division; indeed, the cells may never have left the cell cycle and growth regulator application has some more subtle role. In *Trifolium repens* hypocotyl epidermis, we see that BAP (cytokinin) promotes reorientation of the plane of cell division, leading to initiation of a pro-meristemoid. An analogous response occurs in cotyledon explants of *Abies amabilis* where subepidermal cells develop into shoots. In haploid embryos developed from *Brassica napus* anther cultures, cytokinin actually suppresses secondary embryoid formation and instead promotes normal leafy shoots. This suggests a role for cytokinin in switching between shoot development and embryogenesis. Similarly, Tran Thanh Van *et al.* in1974, working with thin-layer explants only three to six epidermal cells deep from floral branches of *Nicotiana tabacum*, revealed an absolute effect of growth regulators on the direction of differentiation. Although sucrose concentration and light modify the response, structures produced depend mainly on the auxin to cytokinin ratio. At 0.1:1, vegetative shoot buds form and at 100:1 roots are generated, but a 1:1 ratio promotes floral bud initiation. Incidentally, this remains a classic example of formation of new floral meristems *in vitro*.

# Chapter - 3

# Applications of
# Plant Tissue Culture

## Micropropagation

In recent years, the application of micropropagation techniques is employed as an alternative mean of asexual propagation of important plants. It has increased the interest of researchers of biotechnology in diverse fields. The micropropagation techniques are preferred over the conventional asexual propagation methods because of the following reasons : *(a)* in this method only a small amount of tissue is needed as the initial explant for regeneration of millions of clonal plants in a year, *(b)* this method provides a possible alternative method for developing resistance in many species; *(c)* it provides a mean for international exchange of plant materials, hence the problem for introduction of disease can be solved in quarantine; *(d)* *in vitro* stock can be quickly proliferated as it is not season dependent, and *(e)* valuable germplasm can be stored for a long time. Regeneration of plantlets in cultured plant cell and tissues has been achieved in many trees of high economic value. Many of the studies are aimed at large scale micropropagation of important trees yielding fuel, pulp, timber, oils or fruits. Therefore, clonal forestry and horticulture are gaining an increasing recognition as an alternative for tree improvement. However, strategies for transferring cultured plants from *in vitro* to field conditions are based on relatively higher priced horticultural species rather than agricultural and forestry species.

Plant tissue culture studies are being carried out on forest trees worldwide and also in India. Some of the important tree species of interest are: *Acacia nilotica, Albizia lebbeck, A. procera, Azadirachta indica, Bauhinia purpurea, Butea monosperma, Dalbergia* sp., *Dendrocalmus strictus, Eucalyptus* sp., *Ficus religiosa, Morus* sp., *Populus* sp., *Shorea robusta, Tectona grandis (all angiosperms), Biota oriental's, Cedrus deodara, Cryptomena japonica, Picea smithiana, Pinus* sp., *Santalum album* etc.

**PTC** can be used to clone plants and produce many identical plants for a particular market. This can be used when a new variety is grown and other methods of cultivation are too slow for the desired market. It can also be used if a stock plant has been infected and material taken from the plant that is not infected. The excised plant material can be grown on and any disease-free plants grown on for propagation. Plant tissue culture is also of use in research for biochemists, geneticists, plant breeders and plant pathologists. Plant tissue culture has also proved more efficient in the production of secondary metabolites than the use of the parent plants in various instances and has been used in the commercial production of the napthoquinone pigment Shikonin. PTC has also been used in the production of flavours, sweeteners, natural colourants and pharmaceuticals. With the advent of gene insertion plant cells with gene material inserted can be regenerated using tissue culture to produce a whole new plant. Some of the simpler techniques that are more approachable and have been found to be applied directly in plant propagation and genetic improvement of plants are (i) micropropagation, (ii) meristem culture, (iii) somatic embryogenesis, (iv) somaclonal variation, (v) embryo culture, (vi) *in vitro* selection, (vii) anther culture, and (viii) protoplast culture. PTC can be used widely in biotechnology, horticulture, floriculture, agriculture, forestry, pomiculture, production of secondary metabolites of medicinally and commercially importance.

**The most important applications of PTC are as follows;**

1.  Commercial production of plants used in biotechnology, horticulture, floriculture, agriculture, forestry, pomiculture, industry, bioremediation and pharmaceuticals.

2.  For the conservation and cryopreservation of the rare and endangered plant species to avoid extinction.

3.  For the clonal propagation or large-scale production of commercially important plants.

4.  To screen cells, cell aggregates and plantlets for specific characters such as high yield, herbicide resistance/tolerance etc.

5.  Use of meristem tip cultures to produce clean plant material from virus stock such as potatoes.

6.  To produce disease free plants due to its production in sterile environment in large numbers.

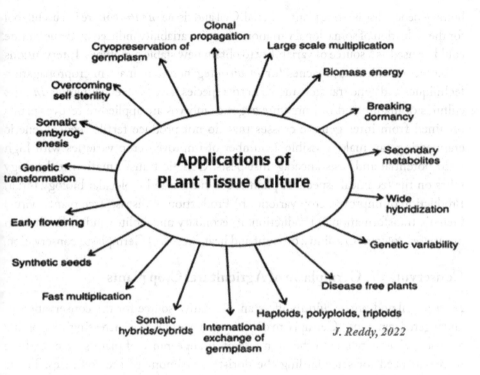

Applications of PLant Tissue Culture

Clonal propagation · Cryopreservation of germplasm · Large scale multiplication · Biomass energy · Overcoming self sterility · Breaking dormancy · Somatic embyrogenesis · Secondary metabolites · Genetic transformation · Wide hybridization · Early flowering · Genetic variability · Synthetic seeds · Disease free plants · Fast multiplication · Somatic hybrids/cybrids · International exchange of germplasm · Haploids, polyploids, triploids

*J. Reddy, 2022*

7. For chromosome doubling and induction of polyploidy for example doubled haploid, tetraploids, and other forms of polyploids.

8. For the development and large-scale production of somatic hybrids and cybrids by protoplast fusion and culture.

9. *In vitro* production of secondary metabolites of industrial and pharmaceutical importance by using hairy root cultures and bioreactors.

10. Development and commercial production of genetically modified plants by *Agrobacterium* mediated biotransformation and Protoplast culture.

### Application of the plant tissue culture in Agriculture

As a well-established technology, PTC has a great impact on both agriculture and industry, through providing plants needed to meet the ever-increasing world demand. It has made significant contributions to the advancement of agricultural sciences in recent times and today they constitute an indispensable tool in modern agriculture Biotechnology has been introduced into agricultural practice at a rate without precedent. Tissue culture allows the production and propagation of genetically homogeneous, disease-free plant material. Cell and tissue *in vitro* culture is a useful tool for the induction of somaclonal variation. Genetic variability induced by tissue culture could be used as a source of variability to obtain new stable genotypes. Interventions of biotechnological approaches for *in vitro* regeneration, mass micropropagation techniques and gene transfer studies in tree species have been encouraging. *In vitro* cultures of mature and/or immature zygotic embryos are applied to recover plants obtained from inter-generic crosses that do not produce fertile seeds. Genetic engineering can make possible a number of improved crop varieties with high yield potential and resistance against pests. Genetic transformation technology relies on the technical aspects of plant tissue culture and molecular biology for; a) Production of improved crop varieties b) Production of disease-free plants (virus) Genetic transformation c) Production of secondary metabolites and d) Production of varieties tolerant to salinity, drought and heat stresses e) Germplasm conservation.

## Conserving the Germplasm of Agricultural Crop plants

*In vitro* cell and organ culture offers an alternative source for the conservation of endangered genotypes. Germplasm conservation worldwide is increasingly becoming an essential activity due to the high rate of disappearance of plant species and the increased need for safeguarding the floristic patrimony of the countries. Tissue culture protocols can be used for preservation of vegetative tissues when the targets for conservation are clones instead of seeds, to keep the genetic background of a crop and to avoid the loss of the conserved patrimony due to natural disasters, whether biotic or abiotic stress. The plant species which do not produce seeds (sterile plants) or which have 'recalcitrant' seeds that cannot be stored for long period of time can successfully be preserved via *in vitro* techniques for the maintenance of gene banks.

Cryopreservation plays a vital role in the long-term *in vitro* conservation of essential biological material and genetic resources. It involves the storage of *in vitro* cells or tissues in liquid nitrogen that results in cryo-injury on the exposure of tissues to physical and chemical stresses. Successful cryopreservation is often ascertained by cell and tissue survival and the ability to re-grow or regenerate into complete plants or form new colonies. It is desirable to assess the genetic integrity of recovered germplasm to determine whether it is 'true-to-type' following cryopreservation. The fidelity of recovered plants can be assessed at phenotypic, histological, cytological, biochemical and molecular levels, although, there are advantages and limitations of the various approaches used to assess genetic stability. Cryobionomics is a new approach to study genetic stability in the cryopreserved plant materials. The embryonic tissues can be cryopreserved for future use or for germplasm conservation.

a.  **Production of rare hybrids:** - Hybridization is a well-established plant breeding procedure to obtain superior plants by combining useful characters distributed in different plants. Hybrid embryo normally aborts on account of the failure of endosperm development or due to embryo endosperm incompatibility. In such instances it can be excised from the young seed & cultured in-vitro. Embryo culture has been successfully applied to several sexually incompatible crosses. In several interspecific crosses of Brassica abortion of hybrid embryo occurs at such an early stage that it is not possible excise & culture the embryo. A hybrid can be obtained by culturing the ovules (ovule culture) or ovaries (ovary culture) enclosing the hybrid embryo. Embryo culture has been used to raise hybrids between the sexually incompatible parents, *Gossypium hirsutum* and *G. arboreum*.

b.  **Somatic hybridization & cybridization:** - Somatic hybridization involves the fusion of somatic cells & regeneration of plants from the fusion products. Plant cells are bounded by a rigid cellulose wall & are cemented together by a pectin-rich matrix to form tissues. An essential step in fusion of plant cells is to bring together the plasma membrane by degrading the cellulosic wall. Thus, the first step in somatic hybridization is the isolation of plant protoplasts. A highly significant application of protoplast fusion is the production of asymmetric hybrids by partial genome transfer from an irradiated donor protoplast to an acceptor protoplast & the selective transfer of cytoplasmic genes. Selective transfer of cytoplasmic traits is achieved by the fusion of normal protoplasts of the recipient parent with the donor's protoplasts in which the nucleus has been rendered inactive by irradiation or with its enucleated such protoplasts or miniplasts. Such hybrids are called cybrids.

c.  **Haploid production:** - Haploids are extremely important in genetics & plant breeding. In haploids it is possible to detect recessive mutants which do not express themselves in diploid state due to the presence of the dominant allele. In cross pollinated plants & F1 hybrids with high degree of heterozygosity, the fixation of a particular trait through the conventional method of backcrossing takes 7-8 years.

## Application of plant tissue culture in Horticulture

a.  **Clonal propagation:** - The conventional method of clonal propagation is slow & often not applicable. For example, the only in-vivo method for clonal multiplication of cultivated orchids, which are complex hybrids is 'back-bulb' propagation. It involves separating the oldest pseudobulbil to force the development buds. This process allows, at best doubling the plant number every year. In-vitro clonal propagation, popularly called Micropropagation.

Micropropagation generally involves three steps: (i) Shoot multiplication (ii) rooting & (iii) transplantation. (i) Shooting multiplication: The most popular method of shoot multiplication is forced proliferation of axillary shoots. For these cultures are initiated from apical or nodal cuttings carrying one or more vegetative buds. In the presence of cytokinin alone or in combination with a low concentration of an auxin, such as IAA or NAA, the pre-existing buds grow & produce 4-6 shoots within 3-4 weeks. (ii) Rooting: Shoots produced through axillary branching or adventitious differentiation are rooted in-vitro on a medium containing a suitable auxin, such as IAA, NAA or IBA. Alternatively, where possible the shoots are treated with auxin & directly planted in potting mixture for in-vivo rooting. (iii) Transplantation: The shoots or plantlets multiplied on a medium containing organic nutrients, show poor photosynthetic capability. In practice, the plants are maintained under high humidity for 10-15 days after they removed from culture vessels. During the next few weeks, the humidity around the plants is gradually lowered, before they are transferred to natural conditions.

b. **Production of disease-free plants:** - Under normal conditions plants are infected by a wide range of pathogens such as bacteria, fungi, viruses, viroids & insects like nematodes & insects. Many perennial plants & those propagated by vegetative means are systematically infected with one or more pathogens, which reduce yield, vigour & quality of the plant. If explants for micropropagation are derived from an infected plant, the pathogens can multiply & spread to a large number of plants. It is therefore essential to use disease free stock plants for micropropagation. Eradication of viruses & other pathogens is also desirable from the point of view of international exchange of plant materials.

**Application of plant tissue culture in Industry:** Cultured plant cells retain their metabolic potential & synthesise secondary products of commerce. Cell cultures can also be used as factories for bioconversion of intermediate compounds into more valuable products. Shikonin, an expensive compound, obtained from the roots of Lithospermum erythrorhizon, has been used by the Japanese traditionally as a vegetable dye & in cosmetics & toiletries. To reduce dependence on import of this plant material, the Japanese scientists have developed a tissue culture method for the commercial production of Shikonin. In tissue cultures the yield of high value compounds can be enhanced by feeding the cells with precursors of their biosynthetic pathway (Biotransformation), manipulation of the culture conditions & selecting high yielding cell lines. One important fundamental contribution of plant tissue culture is the discovery of cytokinins.

**In Vitro Establishment of Mycorrhiza:** Mycorrhizal fungi show highest level of specialization of parasitism. But the major problems with them is their failure to grow on an artificial medium in laboratory. Therefore, establishment and multiplication of mycorrhizal fungi on cultured tissue of the same host plant, if successfully developed, may be a good tool for handling mycorrhizal fungi, production of high potential inoculum and their establishment in root systems of nursery plants in horticulture and forestry, and plantation of mycorrhiza-infested seedlings into field. Only one report is available on this work. Kiernan *et al.* (1984) successfully produced strawberry plants by tissue culture which was infected by a mycorrhizal fungus, *Glomus* sp.

Many attempts have been made to establish Vesicular Arbuscular Mycorrhizal (VAM) fungi in axenic culture but unfortunately none of them got success. It was assumed that self-inhibition of hyphal growth occurs in the growing germ tubes and the self-inhibition compounds were recovered by adding activated charcoal into an agar medium that absorbs inhibitory compounds produced by germ tubes into medium. Cultures of mycorrhizas synthesized aseptically are grouped into two and they are 1. the whole plant cultures and 2. excised root cultures. Both the types of cultures are known as genetobiotic or monogenic systems (Rhodes, 1983). Due to presence of two organisms, it is also known as two-member culture.

Mosse (1962) for the first time, reported the establishment of two member cultures. Appressorium formation and root penetration were much more likely to occur if a *Pseudomonas* sp. was present in culture. It is, therefore, suggested that 3 organisms *i.e.*, fungus-plant-bacterium might be necessary for the development of symbiosis. Moreover, mycorrhizal fungi have been cultured only on cortical tissues of roots which were separated from the whole plant, as in root organ cultures where it acted as food base. VAM fungi have very high degree of specialization for food base on root cortex (Rhodes, 1983). In recent years, after a short period of stagnation, the importance of plants as a source of pharmaceuticals has undergone a resurgence. The reason for this renaissance of plants as a fountain of health and beauty is a sum of many distinct factors. One of these elements is the economic upswing of the Asian countries that has again drawn our attention to the botanical drug market. Traditionally, China is a country where botanical pharma- and nutraceuticals are consumed on a large scale and where these products are fully approved and prescribed, as is also done in many other Asian countries, including Japan and India. A more significant factor contributing to the interest in obtaining new pharmaceuticals from plants is the rapid advancement of molecular technologies. Especially the development of gene transfer methods has facilitated the use of plants as a potential alternative source in the production of pharmaceutical proteins. Recombinant proteins, including vaccines, antibodies, enzymes and regulatory proteins, belong to the rapidly growing sectors in the pharmaceutical industry. The production and expression systems of biologically competitive products should, first of all, be safe and inexpensive to produce. Compared to human and animal cell cultures, plants have several advantages: they are highly scalable, capable of producing biologically active compounds, and free of mammalian viral vectors and pathogens.

## Future Prospects of PTC

The past decades of plant cell biotechnology have evolved as a new era in the field of biotechnology, focusing on the production of a large number of secondary plant products. During the second half of the last century the development of genetic engineering and molecular biology techniques allowed the appearance of improved and new agricultural products which have occupied an increasing demand in the productive systems of several countries worldwide. Nevertheless, these would have been impossible without the development of tissue culture techniques, which provided the tools for the introduction of genetic information into plant cells. Nowadays, one of the most promising methods of producing proteins and other medicinal substances, such as antibodies and vaccines, is the use of transgenic plants. Transgenic plants represent an economical alternative to fermentation-based production systems. Plant-made vaccines or antibodies (plantibodies) are especially striking, as plants are free of human diseases, thus reducing screening costs for viruses and bacterial toxins. The number of farmers who have incorporated transgenic plants into their production systems in 2008 was 13.3 million, in comparison to 11 million in 2007. Examples of currently employed uses of plant tissue culture, especially practical applications of micropropagation will enable agricultural sciences to bring out tremendously potential crops in the years to come. Plant tissue culture industry is estimated to touch US$ 9 billion mark by 2023. Potential uses of plant tissue culture and biotechnology to further our understanding of plant physiology, how plants function and resolution of legal issues will apparently become a serious matter of concern. *Agrobacterium*-mediated genetic transformation along with other modes of transformation such as microprojectile, particle bombardment has been very well accomplished. Incorporation of disease and stress resistance and other horticulturally important traits is a logical outcome to be anticipated in the future. Use of molecular technologies for identification of genotypes, clones and their ancestors will enable researchers and producers to verify identity and parentage of propagules, whether produced by conventional or modern propagation methods. In addition, yet to be imagined applications of plant biotechnology will also emerge as the 21st Century continues to unfold.

# Chapter - 4

# Instruments Used for
# Plant Tissue Culture

Different types of instruments and glass goods are used to culture plant tissues. The conventional and some specific glass goods are required for culture work. Glass goods should be of Corning or Pyrex or similar boro-silicate glass. Measuring cylinder, conical flask, pipettes, beakers are required for preparation of media. Plant tissues are grown in wide-necked Erlenmeyer conical flask (100 ml, 150 ml, 250 ml etc.), culture tubes (25 mm in diameter and 150 mm in length), Petri plates (50, 90, 140 mm in diameter), screw-capped universal bottles (20 cm$^3$ capacity). Sometimes used jam bottle, milk bottle may also be used. Particular care must be taken to ensure that glass goods are properly cleaned before use. The traditional method of cleaning new or dirty glass goods is to soak these in soap water followed by brushing and washing well with tap water and finally rinsing with single distilled water. These are dried in the hot air oven and then the clean glass goods are stored in a dust-proof cupboard or drawer. In order to autoclave the culture medium and to culture the plant material, culture vessels particularly culture flasks and culture tubes must be fitted with cotton plugs which exclude microbial contaminants, yet allow free gas exchange. For this, tightly rolled plugs of non-absorbent cotton wrapped in gauge cloth may be used. When in position the exposed part of each plug and the rim of the culture vessel should be covered by brown paper or a cap of aluminum foil. This will keep the plug and vessel rim free from dust and will protect the plug from wetting during autoclaving. In some laboratories, pre-sterilized, disposable plastic wares are used in order to culture plant tissues. Some of these plastic wares are autoclavable. For the sterilization of medium containing thermo labile compounds or enzymes for protoplast isolation a specially designed glass made bacterial filter or an autoclavable plastic made bacterial filter is used. A small spirit lamp made of glass will be required for the flame sterilization of instruments using methylated spirits. Instruments routinely used for culture work include various sizes of scalpel and forceps, spatula, scissors, etc. All instruments should be of stainless steel. A list of commonly used instruments, their measurement and uses are given above.

## Methods of Sterilization

Asepsis: Plant tissue culture requires contamination free environment, tools and cultures or strict maintenance of germ free system in all the operations, known as asepsis. Particularly in commercial production units, the contamination of one batch of the cultures may result in heavy financial losses or even loss of a culture strain. Therefore strict control measures are enforced to maintain the entry of the personnel and living materials. The basic rules and practices of asepsis are followed in all the tissue culture laboratories.

Plant tissue culture media are rich in nutrients and very suitable for the growth of microbes also. These microorganisms grow faster, consume nutrients rapidly and suppress the growth of plant tissues (by over growth). This saprophytic unwanted growth of microbes is formed as contamination. There are many ways by which these microbes suppress the growth of plant cells and tissues, e.g., release of enzymes and toxins (which inhibit the growth of plant cells), selective absorption of specific nutrients, and some tissue infection.

To achieve success in cultivation of higher plant parts, it is essential to exclude these contaminating microorganisms and hence aseptic techniques must be employed to save cultures. This has led to the development of 'clean area concept'. All the area, tools and working places should be free from microbes to have less and less problems of contamination.

### Following care should be taken:

1. Minimize the air current in the working area so it is possible to avoid spores of contaminating microorganisms to move in along with the air currents over the sterile areas. At least, fan should not be used in laminar air flow bench room (inoculation room). Preferably, all the places should be air-conditioned.

2. Store properly the prepared media, nutrients and tools in cabinets.

3. Use separate area for cleaning and washing and for the preparation of medium.

UV tubes are also fixed in Laminar air flow bench placed in inoculation chamber. All these UV tube lights should be used frequently before inoculations. UV lights of corridors may be left open during nights.

**Surface Sterilization of Explants:** Most contamination is introduced with the explant because of inadequate sterilization or just very dirty material. It can be fungal or bacterial. This kind of contamination can be a very difficult problem when the plant explants material is harvested from the field or greenhouse.

Initial contamination is obvious within a few days after cultures are initiated. Bacteria produce "ooze" on solid medium and turbidity in liquid cultures. Fungi look "furry" on solid medium and often accumulate in little balls in liquid medium.

Bacteria are the most frequent contaminants. They are usually introduced with the explants and may survive surface sterilization of the explants because they are in interior tissues. So, bacterial contamination can first become apparent long after a culture has been initiated. All explants have to be sterilized before transferring them on to the medium.

**Sterilization of the plant material**

**Table: Different disinfectants used for sterilization of plant material**

| Disinfectants | Concentration (%w/v) |
|---|---|
| sodium hypochlorite | 1% |
| Alcohol | 70% |
| hydrogen peroxide | 10% |
| calcium hypochlorite | 7% |
| bromine water | 1% |
| mercuric chloride solution | 0.20% |
| silver nitrate solution | 1% |

It is important to note that while the use of disinfectants decreases the chances of contamination, it increases the possibility of injury to tissues which may hamper the healthy growth of explants. So it is essential to monitor the duration of treatment and to wash away the disinfectant thoroughly after treatment of the explants. Also, some of the disinfectants like bromine water, mercuric chloride and silver nitrate are very harmful for human health. Therefore, disinfectants like sodium hypochlorite and ethanol which are less rigorous are generally preferred. An alternative to avoid the step of disinfection of the explants is to grow the seeds of the desired plant aseptically after sterilization and grow the plants to use them for procuring a healthy, contaminant free explant.

## Method of Ethanol Seed Sterilization

1. Tap seeds into 1.5mL tubes (labeled with VWR pen)
2. Add 1mL 70% Ethanol, 0.1% Triton solution to each tube
3. Vortex at 8 for 5 minutes
4. Aspirate liquid off of seeds
5. Add about 1mL 70% Ethanol to each tube
6. In sterile hood, pipette seeds onto filter paper (labeled with pencil)
7. Let Ethanol evaporate away from filter paper
8. When filter paper is dry, tap seeds onto plates
9. Tape plates, wrap in foil, and refrigerate at 4 degrees for 3-4 days before unwrapping and placing plates in growth chamber

## Suitable sized plant material is sterilized as follows:

1. Clean the working area with ethanol, and start the air flow of the laminar air flow bench.
2. Sterilized petridishes, distilled water, scalpel, filter paper sheets (suitable sized and autoclaved) alcohol, disinfectant (mercuric chloride 0.01-0.1% aqueous solution, or 20% sodium hydrochloride), forceps, bead sterilizer or beaker or coupling jar, sprit lamp or gas burner are collected on Laminar flow bench. Put on the UV light of the bench and put it off after 30 minutes.
3. Bring the plant material to be sterilized on the bench and prepare pieces for sterilization.
4. Clean the working area and hands with alcohol, put on mask and cap, and light the sprit lamp.
5. Keep 3-4 petridishes in a line; add disinfectant in 1st plate and autoclaved distilled water in subsequent plates.
6. Place plant pieces in 1st plate and immerse the material with the help of forceps for 5-10 minutes depending upon the disinfectant used.
7. Transfer material from 1st to 2nd plate, rinse gently and pass to 3rd and 4th plate, one by one with thorough rinsing.
8. Finally, drain the distilled water or place the material in a fresh Petri dish or filter paper and prepare suitable sized explants.

**Mode of Action of Disinfectants:** The various metallic ions can be arranged in a series of decreasing antibacterial activity. $Hg^{2+}$ and $Ag^+$, are effective at less than one part per million (ppm), because of their high affinity for sulphahydryl group. Bacteria are killed by Ag containing $10^5$ to $10^7$ $Ag^+$ ions per cell. The concentration required for killing is markedly affected by inoculum size. Chlorine was the anticeptic introduced (as chlorinated lime) by O.W. Holmes in Boston in 1835 and by Semmelweis in Vienna in 1847, to prevent transmission of puerperal sepsis by the physicians hand. Chlorine combines with water to form hypochlorous acid (HOC1) a strong oxidizing agent.

$$Cl_2 + H_2O = HCl + HOCl$$

$$Cl_2 + 2NaOH = NaCl + NaOCl + H_2O$$

Sodium hypochlorite (NaOCl) solutions are (200 ppm chlorine) are used to sanitize clean surfaces in the food and the dairy industries and in restaurants. Sodium hypochlorite is commercial bleach available in the market as 3-6% solution and can be used directly. It is prepared by passing chlorine in to dilute solution of sodium hydroxide. Powdered bleach is also added into water supply for killing microbes.

**Equipment and Medium Sterilization:** All the manipulations of plant tissue culture methods are carried out in aseptic conditions. All the materials used are therefore, free from microbes or sterilized. Various methods of sterilization are used depending upon the material and type of sterilization.

Essentially, method for sterilization is different for living materials than tools and glass ware. Now a day's many items are available as ready to use, pre-sterilized (by gamma radiation) and disposable. Uses of such disposable materials have facilitated the work.

In laboratory, in principle, three types of sterilization are used:

1. Dry heat
2. Wet heat
3. Filter sterilization.

1. **Dry heat:** Glassware, metal tools and other articles, which do not get charred by high temperature, are put in containers or wrapped in paper or thick aluminum foil and placed in dry oven and sterilized for a period of not less than three hours at a temperature of 140-160 °C. Media and plastic ware cannot be sterilized by this method.

2. **Wet heat:** The most popular method of sterilization both equipment and media is autoclaving at 121°C with a pressure of 15 psi (pounds per square inch) for 15 min (1.02 kg/cm²). Modern autoclaves are capable of providing saturated steam treatment ranging from 70-132 °C, which is a pressure of up to 25 psi. All the vessels containing medium should be placed vertically and should not be filled more than 40% to their total capacity.

3. **Filter sterilization:** Filter sterilization or cold sterilization is used when a solution or medium cannot be sterilized by autoclaving. It is the property of the filter (porosity 0.22 to 0.45 μm) to retain the entire microorganism and make the solution free from microbes. This exclusion of microorganisms makes the solution sterilized without heating or autoclaving.

Only liquids can be sterilized by this method and not the plant materials or other things. This is commonly used where thermolabile or heat sensitive chemicals like plant growth regulators (IAA, GA₃, Zeatin, Abscisic Acid), some bio-chemicals, enzymes for protoplasts isolation, antibiotics or nutrient solution for specific experiment are used to avoid decomposition during autoclaving. After sterilization, solution is added to the medium. In case of static medium, compound is added before solidification (cooling) after autoclaving.

Glass and membrane filters (nitro-cellulose membranes) are commercially available. Glass (sintered glass filters, Borosil) are available in different grades (G-1 to 5) and the G-5 (1-2 pm) is used for bacteriological purposes (Fig. 25.14). These filters have a porous glass disc for filtration of liquids and gases as a filter media which is non-corrosive and reusable.

The grades are classified by maximum pore size which is obtained by measuring the pressure at which the first air bubble breaks away from filter under certain conditions. The pressure differential is then used to calculate the equivalent capillary diameters in microns. The desired pore size is obtained by suitably controlling the grain size, firing time and temperature and the thickness of the disc. These filters are resistant to heat, solvents and detergents.

Membrane type filters (of different diameter) and porosity (0.22 to 0.45 μm) are available as disposable or reusable filters, or as filter holders containing a filter disc (Fig. 25.14). These filters can be connected through tubing or syringe to pass solution or gases. Filters are wrapped in aluminum foil or paper, autoclaved at 121°C for 15 min and then taken to laminar air flow bench. Solution is placed in funnel (glass type filter) or syringe (membrane type filter) and solution is allowed to pass through filter under suction (glass) or pressure (membrane).

A membrane filter or glass tube, filled with cotton may be placed between vacuum flask and suction pump to avoid entry of air from the outside during operation. Sterilized filtrate is collected in flask and dispensed in the liquid or molten medium with the help of sterilized pipette fitted with cotton plug. Volume added to the medium is adjusted to arrive at a final concentration. Membrane type filters are also used at inlets and outlets of various types including air, in a bioreactor system and to draw samples.

There are a variety of wet and dry heat treatments, radiations, filtration and gas and chemical treatments available for direct sterilization of material. Gas treatments are rarely used in the laboratory. Ultraviolet light treatment of working surfaces and sterile rooms are used in the labs while gamma radiation is used in the industry for the preparation of pre-sterilized disposable plastic-wares.

**Sterilization of tools:** Surgical blades and scalpels are not sterilized by dry heat because the high temperature makes the cutting edge dull. Such articles including spatula and forceps are usually immersed in 80% v/v (volume by volume) ethyl alcohol until required, and sterilized during use by frequent immersion in alcohol and flaming. Nowadays, bead sterilizer (works on principle of dry heat) is available for sterilizing such tools. After an initial stabilization time of 30 minutes during which equipment attains a temperature of about 250 °C, units ensure total sterilization by destruction of all micro-organisms within seconds. This unit can conveniently placed on Laminar air-flow bench. Uses of fumigation (formaldehyde, sulphur etc.) have been discouraged. It is used some times to clear the laboratory or working area from microorganisms. The place is not used for 3-4 days following fumigation. Culture rooms are fumigated only after removal of cultures. Generally, wiping of working places with 80% alcohol is sufficient for regular use.

**Sterile transfer area:** An important aspect of maintaining an aseptic environment for tissue culture is the use of sterile transfer area for all the operations of tissue culture. A laminar airflow hood is used to carry out all the steps of tissue culture. All the culture vessels containing autoclaved media or growing plant tissues are opened only while inside the hood and not outside it. This prevents any kind of contamination. Usually tissue culture is carried out in Class II type of hoods which provide an aseptic area to the material inside hood, while the operator remains outside the hood.

A Class II laminar air flow hood consists of:

» Cabinet – Work can be carried out on the floor of the cabinet. The floor is made of stainless steel.

» HEPA (High – efficiency Particulate Air) filter – When air from outside flows into the hood through HEPA filters, it becomes decontaminated as it becomes free of any bacteria.

» Fan – A fan is provided inside the hood, at the top, to maintain positive pressure ventilation. The constant flow of air from the fan prevents the entry of any contaminating bacterial or fungal spores or dust to enter the hood while it is open during working.

» UV Lamp – There is a UV-C germicidal lamp inside the hood. It is switched on for about 15 minutes before starting the work. After UV treatment, the UV lamp is switched off and the hood can be opened to carry out the work.

» Light – A light connection is provided inside the hood for working.

» Gas connection – A gas connection may be provided if gas burners are used. Alternatively, Bunsen burners can be used.

**Autoclaving:** Plant tissue culture media are generally sterilized by autoclaving at 121 °C and 1.05 kg/cm$^2$ (15-20 psi) for 20 minutes. The duration of autoclaving is increased for larger volumes of media. For example, 40 minutes for 1 L media. The time required for sterilization depends upon the volume of medium in the vessel. The minimum times required for sterilization of different volumes of medium are listed below. It is advisable to dispense medium in small aliquots whenever possible as many media components are broken down on prolonged exposure to heat. There is evidence that medium exposed to temperatures in excess of 121 °C may not properly gel or may result in poor cell growth. Several medium components are considered thermolabile and should not be autoclaved. Stock solutions of the heat labile components are prepared and filter sterilized through a 0.22 µm filter into a sterile container. The filtered solution is aseptically added to the culture medium, which has been autoclaved and allowed to cool to approximately 35-45 °C. The medium is then dispensed under sterile conditions. Experimentation with your system is recommended.

Autoclave: Autoclave is used to sterilize medium, glassware and tools for the purpose of plant tissue culture. The same equipment is used in hospitals to sterilize gauge, cotton, tools and linen, etc. Sterilization of material is carried out by increasing moist heat (121 °C) due to increased pressure inside the vessel (15-22 psi, pounds per square inch or 1.02 to 1.5 kg/cm$^2$) for 15 minutes for routine sterilization. Moist heat kills the microorganism and makes the material free from microbes.

**Construction:** Autoclaves of different sizes from 5 litres to several hundred litres capacity are available in horizontal or vertical designs. Autoclaves have a body, an internal (or external for small sized autoclaves comparable to household pressure cooker) heating system, a container to hold material, its cover fixed with pressure gauge, safety valve, pressure release valve etc. Lid is tightened with the help of screws and a gasket seals the body and lid. A jacket, paddle lifter, timer, and indicator etc., are also provided with large sized autoclaves. Autoclaves may be constructed of aluminum, mild steel, stainless steel or gun metal. Industrial autoclave can accommodate large trolley containing huge number of glassware's or large bioreactors.

**Operation**: Place the materials (wrapped in aluminum foil or paper or in metal box) and glass-wares containing medium (plugged with non-absorbent cotton and covered with aluminum foil) in the bucket. Check water level for appropriate level, tighten all the screws, and switch-on the current. Allow the steam to pass freely from release valve for 5 minutes and then close the valve. After attaining a pressure of 15 psi, count 15 minutes for sterilization and then switch-off the current. Pressure is maintained by safety valve. Modern autoclaves are fitted with temperature and time control units and can automatically control the period for sterilization and then switch-off themselves. Empty vessels, beakers, graduated cylinders, etc., should be closed with a cap or aluminum foil. Tools should also be wrapped in foil or paper or put in a covered sterilization tray. It is critical that the steam penetrate the items in order for sterilization to be successful. For large sized vessels and large volume flasks containing high amount of liquid, duration for sterilization should be increased accordingly. The relationship between volume of the solution and duration for its sterilization at 15 psi. The vessels should not be filled more than 1/3 of its capacity for proper sterilization.

**Precautions:**

1.  Check water level each time, the heating elements should remain immersed in the water.

2.  Check spring of safety valve frequently and clean opening whenever necessary.

3.  Opposite screws of the lid should be tightened simultaneously.

4.  Do not over tighten the screws to avoid damage to the gasket.

5.  Use permanent marker to mark your flasks and its medium.

6.  All the electrical equipment should be properly earthen to avoid electric shock.

**Plant Growth Chamber:**

Plant growth chambers can be constructed in a suitable sized room or can be purchased as commercially available equipment. Thermal insulation of walls increases the efficiency of the cooling system.

Essentially plant growth chamber has three environmental control systems:

1. Light-intensity and duration cycle control.

2. Temperature control and regulation.

3. Humidity control and regulation.

All the modern instruments are electronically controlled precision instruments with sophisticated sensors and timers to regulate the desired set values.

**pH Meter:** The hydrogen ion concentration of most solutions is extremely low. In 1909, Sorenson introduced the term pH as a convenient way of expressing hydrogen ion concentration. pH of a solution is strictly defined as the negative logarithm of the hydrogen ion activity. But in practice usually hydrogen ion concentration is taken. $pH = -\log_{10}(H^+) = 7$

The pH of pure water is 7 at 25 °C. Generally glass distilled water is used for the preparation of culture medium. However, sometimes buffered solutions may be used for the same to keep the pH of the medium constant.

**Measurement of pH:** An approximate idea of the pH of a solution can be obtained using indicators. These are organic compounds of natural or synthetic origin whose Colour is dependent upon the pH of the solution. Indicators are usually weak acids, which dissociate in solution. A standard pH meter has two electrodes, one glass electrode for measuring pH and the other calomel reference electrode (Fig. 25.1). Reference electrode is filled with saturated KCl solution. Indicator = Indicator$^-$ + H$^+$. The pH probe measures pH as the activity of hydrogen ions surrounding a thin walled glass bulb at its tip. The probe produces a small voltage (about 0.06 volt per pH unit) that is measured and displayed as pH units by the meter. The meter circuit is fundamentally no more than a voltmeter that displays measurements in pH units instead of volts. The input impedance of the meter must be very high because of the high resistance approximately 20 to 1000 M$\Omega$ of the glass electrode probes typically used with pH meters. The circuit of a simple pH meter usually consists of operational amplifiers in an inverting configuration, with a total voltage gain of about $-17$. The inverting amplifier converts the small voltage produced by the probe (+ 0.059 volt/pH) into pH units, which are then offset by seven volts to give a reading on the pH scale.

For example:

(i) At neutral pH (pH 7) the voltage at the probe's output is 0 volts.

(ii) At alkaline pH, the voltage at the probe's output ranges from + 0 to + 0.41

**Digital balance:** The digital mass balances in the tissue culture labs are very sensitive instruments used for weighing substances to the milligram (0.001 g) level. They need to be treated with lot of care. Use containers when weighing chemicals and always weigh objects at room temperature. Keep the draft shields closed. Do not jar the instruments or change the levels. Always clean the area around the pan with a sable brush after use and inform the stockroom if any liquids or solids spill onto the balance. Always use a container or weighing paper when weighing a chemical; do not place any chemical directly onto the balance pan. Many substances are corrosive and will ruin the sensitive pan and balance mechanisms in just a few minutes. Waxed weighing paper, plastic weighing boats, small beakers, watch glasses, small vials etc. are all convenient containers for weighing chemicals. Additional care must be used when weighing liquids. If possible, flasks containing liquids must be sealed with stoppers to prevent spilling or evaporation during weighing.

**Magnetic stirrer:** **Magnetic mixer** is a laboratory device that employs a rotating magnetic field to cause a stir bar (also called "flea") immersed in a liquid to spin very quickly, thus stirring it. The rotating field may be created either by a rotating magnet or a set of stationary electromagnets, placed beneath the vessel with the liquid.

Magnetic stirrers are often used in chemistry and biology, where they can be used inside hermetically closed vessels or systems, without the need for complicated rotary seals. They are preferred over gear-driven motorized stirrers because they are quieter, more efficient, and have no moving external parts to break or wear out (other than the simple bar magnet itself). Magnetic stir bars work well in glass vessels commonly used for chemical reactions, as glass does not appreciably affect a magnetic field. The limited size of the bar means that magnetic stirrers can only be used for relatively small experiments, of 4 liters or less. Stir bars also have difficulty in dealing with viscous liquids or thick suspensions. For larger volumes or more viscous liquids, some sort of mechanical stirring is typically needed. Because of its small size, a stirring bar is more easily cleaned and sterilized than other stirring devices. They do not require lubricants which could contaminate the reaction vessel and the product. Magnetic stirrers may also include a hot plate or some other means for heating the liquid.

**Shakers:** A shaker is a piece of laboratory equipment used to mix, blend, or to agitate substances in tube(s) or flask(s) by shaking them, which is mainly used in the fields of chemistry and biology. A shaker contains an oscillating board which is used to place the flasks, beakers, test tubes, etc. Although the magnetic stirrer has come to replace the uses of shaker lately, the shaker is still a preferred choice of equipment when dealing with such large volume substances, or simultaneous agitation is required.

## Types of Shakers

**Vortex shaker:** Invented by Jack A. Kraft and Harold D. Kraft in 1962, a Vortex Shaker is usually a small device used to shake, or mix small vials of liquid substance. Its most standout characteristic is that it works by the user putting a vial on the shaking platform and turn it on, thus the vial is shaken along with the platform. Vortex Shaker is vary variable in term of speed adjustment, one can continuously change the shaking speed while shaking by turning the switch.

**Platform shaker:** kind of shaker that has a table board that oscillates horizontally. The liquids to be stirred are held in beakers, jars, or erlenmeyer flasks that are placed over the table; or, sometimes, in test tubes or vials that are nested into holes in the plate.

**Orbital shaker:** Orbital Shaker has a circular shaking motion with a slow speed (25-500 rpm). It is suitable for culturing microbes, washing blots, and general mixing. Some of its characteristics are that they do not create vibrations, and they produce low heat compared to other kinds of shakers, which make it ideal for culturing microbes. Moreover, it can be modified by placing it in an incubator to create an incubator shaker due to its low temperature and vibrations.

**Incubator shaker:** Incubator Shaker (or thermal shaker) can be considered a mix of an incubator and a shaker. It has an ability to shake, while maintain optimal conditions for incubating microbes or DNA replications. This equipment is very useful since in order for a cell to grow, it needs Oxygen, and nutrients, and that require shaking so that they can be distributed evenly around the culture.

**Controlling light:** Light is fixed in the roof of equipment or in shelves. Light is provided by commercially available light sources like cool white fluorescent tubes and incandescent lamps in a ratio of 3:1 and usually a light intensity of 2000-2500 lux (about 200-250 candles or 30 $\mu$ mol m$^{-2}$ s$^{1}$) is used. The duration of light and dark cycle is adjusted as per requirement, usually 16 hours light cycle is given. Nowadays warm fluorescent tubes are also available which provides wide spectrum as compared to cool white fluorescent tubes and mixing of incandescent light is not required with former tubes. Light intensity can also be regulated by photoperiod simulators. Thus, light quality, intensity and period is controlled and regulated by the instrument as per set valves.

**Maintenance of temperature**: In modern equipment, temperature is precisely regulated by good quality (platinum) temperature sensor. In all cases, air conditioning units provide the cooling. It is always advisable to keep one spare compressor unit, for emergency, to avoid delay in repairs and damage to cultures. Usually temperature of 22-28 °C is used for growing plant tissue culture. Temperatures should be measured in a constructed growth chamber at different levels and places, viz., light racks, central and corners to have a correct temperature setting.

**Maintenance of humidity**: Humidity inside the growth chamber is provided by humidifier (a mist generating system) and controlled by humidistat. Usually 60% RH (relative humidity) is used to maintain healthy growth. Low RH may cause early drying of medium while high humidity may cause fungal growth in the environment and on a various articles. Thus, in a growth chamber, light, temperature and humidity are precisely controlled and cultures are grown in a controlled environment. All the controls are set on control panel.

**Microscopy:**

**(a) Electron Microscopy**: Electron microscopy permits a detailed study of sub-cellular organelles as its resolving power is much greater than that of the light microscope. Max Knoll and Ernst Ruska in 1931, at Technical University in Berlin, constructed electron microscope (EM). In the EM, streams of electrons are deflected by an electrostatic or electromagnetic field in the same way that a beam of light is refracted by a lens. Electron beam is generated by heating a filament in vacuum, which are accelerated by a potential and shows properties similar to light ($\lambda$ = 0.005nm of electrons and 550 nm for light). Though appears to be similar there are great differences in light and electron microscopes, the principal being the image formation (Fig. 25.4.). In electron microscope, image is produced by electron scattering. Electron dispersion is a function of the thickness and molecular packing of the object and depends especially on the atomic number of the atoms in the object.

The resolving power of Transmission Electron Microscope (TEM) is very high. The image generated by objective can be multiplied several hundred times by projector coil, e.g., 100 objective 200x projector coil = 20,000x. In TEM, this can be reached up to 10,000,000 xs. Electron microscope has a greater depth of field as compared to light microscope.

**(b) SEM:** Scanning electron microscope (SEM) provides surface views of whole structure of specimen (Fig. 25.7). Normally, specimens are coated with a thin film of metal under vacuum. As compared to shadow casting, complete coating is essential because the scanning beam produces a charge on the uncoated biological materials which cause distortion of image. The metal coated specimen is scanned with a beam of electrons (50-100 Å) which strikes on the specimen and emits the secondary electrons. These electrons pass through a cathode tube and image is produced on the screen of Cathode Ray Tube. In plant system, SEM is used to see the nature of appendages, pollen morphology and morphogenesis in plant tissue culture. It is also used to study the structures of metals, crystals and in forensic sciences.

**(c) Light Microscopy:** Bright field microscopy is absolutely indispensable tool for cell biologists. This is required for routine observations of cells, cellular differentiation and pigmentation. Fluorescence microscopy has become a powerful tool for cell biologists, particularly for the selection of fluorescing secondary metabolite rich cells. Only these two techniques are discussed here in brief. According to wave theory, light is propagated from one place to another as wave travelling in a hypothetical medium. Light waves are described in terms of their amplitude, frequency, and wavelength. Amplitude is the maximum displacement of light path from the position of equilibrium. Frequency of light is the number of complete cycles occurring in a second. The property of light waves inversely associated with the frequency is the wavelength and is defined as the distance between corresponding points on a wave or distance between two successive peaks or crests. The velocity of light is about 186300 miles or $3 \times 10^{15}$ kms per second. A good microscope has not only good magnifying power but also good resolving power to provide finer details of the object. Thus, the basic difficulty in designing a microscope is riot the magnification, but the ability of lens system to distinguish two adjacent points as distinct and separate. This ability is known as resolving power of the microscope. The resolving power of a microscope depends upon the wavelength of light and numerical aperture (N.A.).

**The minimum resolvable distance between two luminous points (v) is given by the following formula:**

$$V = \frac{0.16\lambda}{N.A.} \quad \text{where, } \lambda = \text{wavelength}$$

Thus, shorter the wavelength of light used and lower the N.A., greater is the resolving power. The limit of resolution of a microscope is approximately equal to 0.5 /N.A., which for a light microscope is approximately 200 nanometers (nm) or about the size of many bacterial cells. The numerical aperture of a lens is dependent on the refractive index (r) = the ratio of the speed of light in a given medium to the speed of light in a vacuum) of the medium filling the space between the specimen and the front of the objective lens and on the angle of the most oblique rays of light that can enter the objective lens ($\theta$). It is given by formula [N.A. = $\eta x \sin\theta$], that is why immersion oil is placed between the object and oil immersion lens (100 X objective).

**Colorimeter:** The most commonly used method for determining the concentration of biochemical compounds is colorimetry. It uses the property of light such that when white light passes through a coloured solution, some wavelengths are absorbed more than others. Hyaline solution can be made coloured by specific reactions with suitable reagents. These reactions are generally very sensitive to determine quantities of material in the region of millimole per litre concentration. The big advantage is that complete isolation of the compound is not necessary and the constituents of a complex mixture such as blood can be determined after little treatment. The depth of colour is directly proportional to the concentration of the compound being measured, while the amount of light absorbed is proportional to the intensity of the colour and therefore, to the concentration.

**Centrifugation:** A centrifuge is an instrument which produces centrifugal force by rotating the samples around a central axis with the help of an electric motor. Centrifuges can be categorized as the clinical type (5-10,000 rpm), refrigerated high-speed centrifuges (10,000-20,000 rpm) and ultra -centrifuges (20,000 to 80,000 rpm). With increase in rpm, the friction of rotor with air produces so much of heat that they have to be run under refrigeration (so called refrigerated centrifuge) and both refrigeration and vacuum are used in ultracentrifuge, which runs at very high rpm. For these high speeds, even the rotor has to be made of special metal to withstand the great force. There are two types of rotors, angle head and swing (Fig. 25.10). In the former, the samples are kept at an angle of about 30° to the vertical axis whereas; in the latter the samples while spinning are horizontal. Simple calculations show that for the same radius, the swinging bucket method produces more gravitational force. Ultracentrifuges are of two types – analytical and preparative model.

**Analytical model:** This consists of rotors and tubes, called cells. The instrument is designed to allow the operator to follow the progress of the substances in the cells, while the process of centrifugation is in progress. By estimating sedimentation velocity during the process, the molecular weight, purity etc. can be determined.

**Preparative model:** This is used for purification of the components of macromolecules or other substances and all determinations are made at the end of centrifugation.

The instrument has no monitoring device, while large centrifugal forces are set for a fixed time period. Centrifugation is the most widely used technique for separation of various metabolites and also used to separate non-miscible liquids during extraction of secondary metabolites, e.g., water (aqueous), chloroform (organic) mixture.

A wide variety of centrifuges are available, ranging in capacity and speed. During the process of centrifugation, solid particles experience a centrifugal force, which pulls them outwards, i.e., away from the center. The velocity with which a given solid particle moves through a liquid medium is related to angular velocity. The principle of centrifugation is that, an object moving in a circular motion at an angular velocity is subjected to an outward force (F) through a radius of rotation (r) in cms, presented as $\omega^2[F = \omega^2 r]$.

F frequently expressed in terms of gravitational force of the Earth, commonly referred to as RCF (Relative centrifugal force) or g by the following formula [$RCF = \omega 2r / 980$]. The $\omega^2$ operating speed of the centrifuge is expressed as revolutions per minute 'rpm', which can be converted to radians by the following formula:

$$\omega = \frac{rpm}{30}\frac{\pi\,(rpm)}{30}$$

or,
$$RCF = \frac{(\pi\,rpm)^2\,(r)}{\dfrac{30^2}{980}} = \frac{980\,(\pi\,rpm)^2\,(r)}{30^2}$$

Therefore,
$$RCF = 1.2 \times 10^{-5}\,(rpm)^2\,r$$

Sometimes the velocity of the moving particles is expressed in the form of sedimentation coefficient (s) by the formula [$V - s\,(\omega^2 r)$]. The sedimentation coefficient is a characteristic constant for a molecule or a particle and is a function of the size, shape and density. It is equivalent to the average velocity per unit of acceleration. The unit Svedberg (s) is often used with reference to centrifugation and is equivalent to a sedimentation coefficient of $10^{-13}$ s.

**Thermometer:** The thermometer is a device that measures temperature or temperature gradient using a variety of different principles; it comes from the Greek roots thermo, heat, and meter, to measure. A thermometer has two important elements: the temperature sensor (e.g., the bulb on a mercury thermometer) in which some physical change occurs with temperature, plus some means of converting this physical change into a value (e.g., the scale on a mercury thermometer). Industrial thermometers commonly use electronic means to provide a digital display or input to a computer. The Alcohol thermometer or Spirit thermometer is an alternative to the Mercury-in-glass thermometer, and functions in a similar way. An organic liquid is contained in a glass bulb which is connected to a capillary of the same glass and the end is sealed with an expansion bulb. The space above the liquid is a mixture of nitrogen and the vapour of the liquid.

For the working temperature range, the meniscus or interface between the liquid is within the capillary. With increasing temperature, the volume of liquid expands and the meniscus moves up the capillary. The position of the meniscus shows the temperature against an inscribed scale.

The liquid used can be pure ethanol or toluene or kerosene or Isoamyl acetate, depending on manufacturer and working temperature range. Since these are transparent, the liquid is made more visible by the addition of a red or blue dye. One half of the glass containing the capillary is usually enameled white or yellow to give a background for reading the scale. Temperature is measured by maximum-minimum thermometer or a continuous rotary dram chart type thermometer. The U- shaped maximum-minimum thermometer is commonly used for determining diurnal maximum and minimum range of temperature in the culture room. During the dark period, temperature remains slightly (1-2 °C) lower than light period (all bulbs and tube lights of the culture room increase the temperature). Therefore, chokes of the tube lights are fitted outside the culture room. The indicators of the maximum-minimum thermometer are moved by mercury column and they remain at that position until moved by the observer with the help of magnet. Their positions indicate the minimum and maximum temperature in the previous 24 h. After recording temperature indicators are reset to mercury level.

**Hygrometer:** Hygrometers are instruments used for measuring humidity. A simple form of a hygrometer is specifically known as a "psychrometer" and consists of two thermometers, one of which includes a dry bulb and the other of which includes a bulb that is kept wet to measure wet-bulb temperature. Evaporation from the wet bulb lowers the temperature, so that the wet-bulb thermometer usually shows a lower temperature than that of the dry-bulb thermometer, which measures dry-bulb temperature. Relative humidity is computed from the ambient temperature as shown by the dry- bulb thermometer and the difference in temperatures as shown by the wet-bulb and dry-bulb thermometers. Relative humidity can also be determined by locating the intersection of the wet- and dry-bulb temperatures on a psychrometric chart. Dial type hair hygrometer is a convenient tool to measure humidity in the culture room. Humidifier is used to maintain 60% relative humidity (RH) in the culture room. Humidistat is a bimetallic thermocouple device to control and regulate the function of humidifier to maintain the humidity. Distilled water should be filled in the humidifier. The RH present in the culture room is measured by hair hygrometer. As the name suggests a chemically treated hair elongates with increased humidity and shortens with dryness (similar to mercury in thermometer). It is calibrated from 0 to 100% RH. At lower humidity, medium dries rapidly whereas at higher humidity chances of fungal growth over all surfaces and cotton plugs is increased. Therefore, about 60% RH is maintained in the culture room.

**LUX meter:** Light is the form of radiant energy, i.e., electromagnetic radiation of specific wavelength. Visible light as we perceive, is located in narrow wavelength region of spectrum between 380 to 760 nm. Light has dual characters, displaying both wave properties (refraction, diffraction, interference and polarization phenomena) and particle properties (light is radiated in discrete amounts of energy or photons). Properties of light should be defined either as irradiance (radiant flux intercepted per unit area; unit = w/m$^2$) or an illuminance (luminous flux intercepted per unit area, unit lux, 10.76 lux = 1 foot candle). It is to note that an irradiance measurement is not spectrally defined, whereas, an illuminance measurement indicates the level of visible light as the human eye would see it.

Artificial light is provided by cool, white, fluorescent tubes and/or incandescent bulbs of different classes of artificial light sources, fluorescent sources have been used almost exclusively for plant tissue culture because they are more efficient at producing broad band visible light than incandescent bulbs and are available in low output wattage than lamps.

Light intensity can be measured by photometer or lux meter. A photometer consists of photoelectric cell and a micro-ammeter. Photoelectric cell is sensitive for light and converts light into current. Micro-ammeter shows reading due to this current and its needle moves. Nowadays digital read out is given by appropriately converting the current into digital signal. High intensity is proportional to the current generated in the photoelectric cell by falling light. The unit of illumination is called as 'lux' and the scale of ammeter is calibrated in lux. Usually 2000 to 2500 lux is provided to the cultures maintained in light in the culture room. The instrument is a sensitive tool and handled with care. It should not be exposed to the sunlight without switching the proper reading switch to high light illumination.

# Chapter - 5

# Plant Tissue Culture
# Laboratory Organisation

## Instruments Used for Plant Tissue Culture

The conventional and some specific glass goods are required for culture work. Glass goods should be of Corning or Pyrex or similar boro-silicate glass. Measuring cylinder, conical flask, pipettes, beakers are required for preparation of media. Plant tissues are grown in wide-necked Erlenmeyer conical flask (100 ml, 150 ml, 250 ml etc.), culture tubes (25 mm in diameter and 150 mm in length), Petri plates (50, 90, 140 mm in diameter), screw-capped universal bottles (20 cm$^3$ capacity). Sometimes used jam bottle, milk bottle may also be used. Particular care must be taken to ensure that glass goods are properly cleaned before use. The traditional method of cleaning new or dirty glass goods is to soak these in soap water followed by brushing and washing well with tap water and finally rinsing with single distilled water. These are dried in the hot air oven and then the clean glass goods are stored in a dust-proof cupboard or drawer. In order to autoclave the culture medium and to culture the plant material, culture vessels particularly culture flasks and culture tubes must be fitted with cotton plugs which exclude microbial contaminants, yet allow free gas exchange. For this, tightly rolled plugs of non-absorbent cotton wrapped in gauge cloth may be used. When in position the exposed part of each plug and the rim of the culture vessel should be covered by brown paper or a cap of aluminum foil. This will keep the plug and vessel rim free from dust and will protect the plug from wetting during autoclaving. In some laboratories, pre-sterilized, disposable plastic wares are used in order to culture plant tissues. Some of these plastic wares are autoclavable. For the sterilization of medium containing thermo labile compounds or enzymes for protoplast isolation a specially designed glass made bacterial filter or an autoclavable plastic made bacterial filter is used. A small spirit lamp made of glass will be required for the flame sterilization of instruments using methylated spirits. Instruments routinely used for culture work include various sizes of scalpel and forceps, spatula, scissors, etc. All instruments should be of stainless steel. A list of commonly used instruments, their measurement and uses are given below in this chapter.

**Organization of the laboratory:** For the culture of plant cell and tissue, you require a well-equipped laboratory. Laboratory set-up depends on the nature of research to be carried out and availability of funds. But the most basic facility that an individual needs for tissue culture requires the following:

i.    **Washing and storage facilities:** A separate area is required which should have large sink with provision for hot and cold running water, distillation apparatus, washing machine, pipette washer, drier and cleaning brushes.

ii.   **Media preparation room:** An area is required for preparation of media. In such space there should be provision for bench space for chemicals, labware, culture vessels, closures and miscellaneous equipment required for media preparation and dispensing. In this room provision is also made for placing hot plates or stirrers, pH meter, balance, water bath, burners, oven, autoclave, culture vessel, refrigerator, etc.

**(iii) Inoculation chamber/Transfer area:** All the activities of sterile transfers are performed in this room. There must be a laminar air flow cabinet where all the precautions should be taken to prevent entry of any contaminant into the culture vial during the process of inoculation or subculture. Laminar air flow hoods are usually sterilized by switching on the hood and wiping the working surface with 70% ethyl alcohol for 15 minutes before initiating any operation under the hood. Ultraviolet light (UV) is sometimes installed to disinfect the area; this light should only be used when people and plant materials are not in the room. This room is provided with:

1.    Laminar air flow cabinet: Inoculation at subculture by maintaining aseptic condition.

2.    Steri-bed sterilizer, Sprit lamp/Bunsen burner: Sterilization of the knives, scalpels, forceps etc.

3.    Stereo-microscope: observe for specific part.

4.    Ethyl alcohol: sterilization and flaming of small instruments.

5.    Tiles/glass plates use during sterile cutting.

6.    Hypochlorite solution: sterilization of plant material

7.  Kitchen timer: timing for sterilization.

**(iv) Culture room:** This is the room where light, temperature, humidity are maintained. All of these environmental considerations will vary depending on the size of the growth room. Temperature is an important consideration for the tissue culture and other factors like light, relative humidity, and shelving depend on it. Generally, temperature of the growth room remains in the range of 25± 2°C. Temperature in the primary growth room can be maintained by air conditioner.

**Lighting facility:** Intensity of light in the room can easily be maintained by using fluorescent light with timer. However, most culture rooms are lighted at the 1000 lux (for 1000cft) with some going up 5000-10000 lux.

**Light duration:** 16-18 h/day.

**Light quality:** Spectral quality of light received by in vitro cultures is very important.

**Relative humidity:** Relative humidity (RH) is very difficult to control inside the room but humidifier can be used to control humidity. Humidity inside the room should be 70-75%

**Shelves:** Shelving with primary growth rooms can vary depending upon the situations & explants grown. Wood is recommended for the inexpensive easy to build shelves.

This room is provided with

1.  Temperature control (25± 2°C)

2.  Electricity supply essential for lighting, cooling and heating

3.  Shelves for culture racks

4.  Fluorescent tubes for lighting

5.  Timer for regulating day length

6.  Racks for culture vials

7.  Rotary shaker for suspension cultures

8.  Observations table.

**(v) Data Collection (Observation):** The cultures are monitored at regular intervals in the culture room for the growth and development of cultured tissues. Observation is also made under aseptic area in laminar airflow.

**(vi) Acclimatization area:** Plants regenerated from in vitro tissue cultures are transplanted to vermiculite pots. The potted plants are ultimately transferred to greenhouses or growth cabinets and maintained for further observations under controlled conditions of light, temperature and humidity.

(vi) Additional facilities:

> » Photoperiodic simulation castor racks
> » Laminar air flow with latest ozone ionization disinfection
> » Heat convectors
> » Air-conditioning
> » Humidity monitor microclimate
> » Floor insulation polymer anti-conductive Buffer room
> » Photoperiodic simulation morphogen
> » Temperature controller microclimate
> » Ceiling insulation polyurethane
> » Air curtains etc.

**Sterile transfer area:** An important aspect of maintaining an aseptic environment for tissue culture is the use of sterile transfer area for all the operations of tissue culture. A laminar airflow hood is used to carry out all the steps of tissue culture. All the culture vessels containing autoclaved media or growing plant tissues are opened only while inside the hood and not outside it. This prevents any kind of contamination. Usually tissue culture is carried out in Class II type of hoods which provide an aseptic area to the material inside hood, while the operator remains outside the hood.

A Class II laminar air flow hood consists of:

> » Cabinet – Work can be carried out on the floor of the cabinet. The floor is made of stainless steel.
> » HEPA (High – efficiency Particulate Air) filter – When air from outside flows into the hood through HEPA filters, it becomes decontaminated as it becomes free of any bacteria.

» Fan – A fan is provided inside the hood, at the top, to maintain positive pressure ventilation. The constant flow of air from the fan prevents the entry of any contaminating bacterial or fungal spores or dust to enter the hood while it is open during working.

» UV Lamp – There is a UV-C germicidal lamp inside the hood. It is switched on for about 15 minutes before starting the work. After UV treatment, the UV lamp is switched off and the hood can be opened to carry out the work.

**Autoclaving:** Plant tissue culture media are generally sterilized by autoclaving at 121 °C and 1.05 kg/cm² (15-20 psi) for 20 minutes. The duration of autoclaving is increased for larger volumes of media. For example, 40 minutes for 1 L media. The time required for sterilization depends upon the volume of medium in the vessel. The minimum times required for sterilization of different volumes of medium are listed below. It is advisable to dispense medium in small aliquots whenever possible as many media components are broken down on prolonged exposure to heat. There is evidence that medium exposed to temperatures in excess of 121 °C may not properly gel or may result in poor cell growth. Several medium components are considered thermolabile and should not be autoclaved. Stock solutions of the heat labile components are prepared and filter sterilized through a 0.22 µm filter into a sterile container. The filtered solution is aseptically added to the culture medium, which has been autoclaved and allowed to cool to approximately 35-45 °C. The medium is then dispensed under sterile conditions. Experimentation with your system is recommended.

**Autoclave:** Autoclave is used to sterilize medium, glassware and tools for the purpose of plant tissue culture. The same equipment is used in hospitals to sterilize gauge, cotton, tools and linen, etc. Sterilization of material is carried out by increasing moist heat (121 °C) due to increased pressure inside the vessel (15-22 psi, pounds per square inch or 1.02 to 1.5 kg/cm²) for 15 minutes for routine sterilization. Moist heat kills the microorganism and makes the material free from microbes.

**Construction:** Autoclaves of different sizes from 5 litres to several hundred litres capacity are available in horizontal or vertical designs. Autoclaves have a body, an internal (or external for small sized autoclaves comparable to household pressure cooker) heating system, a container to hold material, its cover fixed with pressure gauge, safety valve, pressure release valve etc. Lid is tightened with the help of screws and a gasket seals the body and lid. A jacket, paddle lifter, timer, and indicator etc., are also provided with large sized autoclaves. Autoclaves may be constructed of aluminum, mild steel, stainless steel or gun metal. Industrial autoclave can accommodate large trolley containing huge number of glassware's or large bioreactors.

**Operation**: Place the materials (wrapped in aluminum foil or paper or in metal box) and glass-wares containing medium (plugged with non-absorbent cotton and covered with aluminum foil) in the bucket. Check water level for appropriate level, tighten all the screws, and switch-on the current. Allow the steam to pass freely from release valve for 5 minutes and then close the valve. After attaining a pressure of 15 psi, count 15 minutes for sterilization and then switch-off the current. Pressure is maintained by safety valve. Modern autoclaves are fitted with temperature and time control units and can automatically control the period for sterilization and then switch-off themselves. Empty vessels, beakers, graduated cylinders, etc., should be closed with a cap or aluminum foil. Tools should also be wrapped in foil or paper or put in a covered sterilization tray. It is critical that the steam penetrate the items in order for sterilization to be successful. For large sized vessels and large volume flasks containing high amount of liquid, duration for sterilization should be increased accordingly. The relationship between volume of the solution and duration for its sterilization at 15 psi. The vessels should not be filled more than 1/3 of its capacity for proper sterilization.

**Precautions:**

1. Check water level each time, the heating elements should remain immersed in the water.

2. Check spring of safety valve frequently and clean opening whenever necessary.

3. Opposite screws of the lid should be tightened simultaneously.

4. Do not over tighten the screws to avoid damage to the gasket.

5. Use permanent marker to mark your flasks and its medium.

6. All the electrical equipment should be properly earthen to avoid electric shock.

## 3. Plant Growth Chamber

Plant growth chambers can be constructed in a suitable sized room or can be purchased as commercially available equipment. Thermal insulation of walls increases the efficiency of the cooling system.

Essentially plant growth chamber has three environmental control systems:

1. Light-intensity and duration cycle control.

2. Temperature control and regulation.

3. Humidity control and regulation.

All the modern instruments are electronically controlled precision instruments with sophisticated sensors and timers to regulate the desired set values.

**pH Meter:** The hydrogen ion concentration of most solutions is extremely low. In 1909, Sorenson introduced the term pH as a convenient way of expressing hydrogen ion concentration. pH of a solution is strictly defined as the negative logarithm of the hydrogen ion activity. But in practice usually hydrogen ion concentration is taken. $pH = -\log_{10} (H^+) = 7$

The pH of pure water is 7 at 25 °C. Generally, glass distilled water is used for the preparation of culture medium. However, sometimes buffered solutions may be used for the same to keep the pH of the medium constant.

**Measurement of pH:** An approximate idea of the pH of a solution can be obtained using indicators. These are organic compounds of natural or synthetic origin whose Colour is dependent upon the pH of the solution. Indicators are usually weak acids, which dissociate in solution. A standard pH meter has two electrodes, one glass electrode for measuring pH and the other calomel reference electrode (Fig. 25.1). Reference electrode is filled with saturated KCl solution. Indicator = Indicator$^-$ + $H^+$. The pH probe measures pH as the activity of hydrogen ions surrounding a thin-walled glass bulb at its tip. The probe produces a small voltage (about 0.06 volt per pH unit) that is measured and displayed as pH units by the meter. The meter circuit is fundamentally no more than a voltmeter that displays measurements in pH units instead of volts. The input impedance of the meter must be very high because of the high resistance approximately 20 to 1000 M$\Omega$ of the glass electrode probes typically used with pH meters. The circuit of a simple pH meter usually consists of operational amplifiers in an inverting configuration, with a total voltage gain of about − 17. The inverting amplifier converts the small voltage produced by the probe (+ 0.059 volt/pH) into pH units, which are then offset by seven volts to give a reading on the pH scale.

For example:

(i) At neutral pH (pH 7) the voltage at the probe's output is 0 volts.

(ii) At alkaline pH, the voltage at the probe's output ranges from + 0 to + 0.41

**Digital balance:** The digital mass balances in the tissue culture labs are very sensitive instruments used for weighing substances to the milligram (0.001 g) level. They need to be treated with lot of care. Use containers when weighing chemicals and always weigh objects at room temperature. Keep the draft shields closed. Do not jar the instruments or change the levels. Always clean the area around the pan with a sable brush after use and inform the stockroom if any liquids or solids spill onto the balance. Always use a container or weighing paper when weighing a chemical; do not place any chemical directly onto the balance pan. Many substances are corrosive and will ruin the sensitive pan and balance mechanisms in just a few minutes. Waxed weighing paper, plastic weighing boats, small beakers, watch glasses, small vials etc. are all convenient containers for weighing chemicals. Additional care must be used when weighing liquids. If possible, flasks containing liquids must be sealed with stoppers to prevent spilling or evaporation during weighing.

**Magnetic stirrer:** **Magnetic mixer** is a laboratory device that employs a rotating magnetic field to cause a stir bar (also called "flea") immersed in a liquid to spin very quickly, thus stirring it. The rotating field may be created either by a rotating magnet or a set of stationary electromagnets, placed beneath the vessel with the liquid.

Magnetic stirrers are often used in chemistry and biology, where they can be used inside hermetically closed vessels or systems, without the need for complicated rotary seals. They are preferred over gear-driven motorized stirrers because they are quieter, more efficient, and have no moving external parts to break or wear out (other than the simple bar magnet itself). Magnetic stir bars work well in glass vessels commonly used for chemical reactions, as glass does not appreciably affect a magnetic field. The limited size of the bar means that magnetic stirrers can only be used for relatively small experiments, of 4 liters or less. Stir bars also have difficulty in dealing with viscous liquids or thick suspensions. For larger volumes or more viscous liquids, some sort of mechanical stirring is typically needed. Because of its small size, a stirring bar is more easily cleaned and sterilized than other stirring devices. They do not require lubricants which could contaminate the reaction vessel and the product. Magnetic stirrers may also include a hot plate or some other means for heating the liquid.

**Shakers:** A shaker is a piece of laboratory equipment used to mix, blend, or to agitate substances in tube(s) or flask(s) by shaking them, which is mainly used in the fields of chemistry and biology. A shaker contains an oscillating board which is used to place the flasks, beakers, test tubes, etc. Although the magnetic stirrer has come to replace the uses of shaker lately, the shaker is still a preferred choice of equipment when dealing with such large volume substances, or simultaneous agitation is required.

## Types of Shakers

**Vortex shaker:** Invented by Jack A. Kraft and Harold D. Kraft in 1962, a Vortex Shaker is usually a small device used to shake, or mix small vials of liquid substance. Its most standout characteristic is that it works by the user putting a vial on the shaking platform and turn it on, thus the vial is shaken along with the platform. Vortex Shaker is vary variable in term of speed adjustment, one can continuously change the shaking speed while shaking by turning the switch.

**Platform shaker:** kind of shaker that has a table board that oscillates horizontally. The liquids to be stirred are held in beakers, jars, or erlenmeyer flasks that are placed over the table; or, sometimes, in test tubes or vials that are nested into holes in the plate.

**Orbital shaker:** Orbital Shaker has a circular shaking motion with a slow speed (25-500 rpm). It is suitable for culturing microbes, washing blots, and general mixing. Some of its characteristics are that they do not create vibrations, and they produce low heat compared to other kinds of shakers, which make it ideal for culturing microbes. Moreover, it can be modified by placing it in an incubator to create an incubator shaker due to its low temperature and vibrations.

**Incubator shaker:** Incubator Shaker (or thermal shaker) can be considered a mix of an incubator and a shaker. It has an ability to shake, while maintain optimal conditions for incubating microbes or DNA replications. This equipment is very useful since in order for a cell to grow, it needs Oxygen, and nutrients, and that require shaking so that they can be distributed evenly around the culture.

**Controlling light:** Light is fixed in the roof of equipment or in shelves. Light is provided by commercially available light sources like cool white fluorescent tubes and incandescent lamps in a ratio of 3:1 and usually a light intensity of 2000-2500 lux (about 200-250 candles or 30 $\mu$ mol m$^{-2}$ s$^1$) is used. The duration of light and dark cycle is adjusted as per requirement, usually 16 hours light cycle is given. Nowadays warm fluorescent tubes are also available which provides wide spectrum as compared to cool white fluorescent tubes and mixing of incandescent light is not required with former tubes. Light intensity can also be regulated by photoperiod simulators. Thus, light quality, intensity and period is controlled and regulated by the instrument as per set valves.

**Maintenance of temperature:** In modern equipment, temperature is precisely regulated by good quality (platinum) temperature sensor. In all cases, air conditioning units provide the cooling. It is always advisable to keep one spare compressor unit, for emergency, to avoid delay in repairs and damage to cultures. Usually, temperature of

22-28 °C is used for growing plant tissue culture. Temperatures should be measured in a constructed growth chamber at different levels and places, viz., light racks, central and corners to have a correct temperature setting.

**Maintenance of humidity:** Humidity inside the growth chamber is provided by humidifier (a mist generating system) and controlled by humidistat. Usually 60% RH (relative humidity) is used to maintain healthy growth. Low RH may cause early drying of medium while high humidity may cause fungal growth in the environment and on a various article. Thus, in a growth chamber, light, temperature and humidity are precisely controlled and cultures are grown in a controlled environment. All the controls are set on control panel.

## Microscopy

**(a) Electron Microscopy:** Electron microscopy permits a detailed study of sub-cellular organelles as its resolving power is much greater than that of the light microscope. Max Knoll and Ernst Ruska in 1931, at Technical University in Berlin, constructed electron microscope (EM). In the EM, streams of electrons are deflected by an electrostatic or electromagnetic field in the same way that a beam of light is refracted by a lens. Electron beam is generated by heating a filament in vacuum, which are accelerated by a potential and shows properties similar to light ($\lambda$ = 0.005nm of electrons and 550 nm for light). Though appears to be similar there are great differences in light and electron microscopes, the principal being the image formation (Fig. 25.4.). In electron microscope, image is produced by electron scattering. Electron dispersion is a function of the thickness and molecular packing of the object and depends especially on the atomic number of the atoms in the object.

The resolving power of Transmission Electron Microscope (TEM) is very high. The image generated by objective can be multiplied several hundred times by projector coil, e.g., 100 objective 200x projector coil = 20,000x. In TEM, this can be reached up to 10,000,000 times. Electron microscope has a greater depth of field as compared to light microscope.

**(b) SEM:** Scanning electron microscope (SEM) provides surface views of whole structure of specimen (Fig. 25.7). Normally, specimens are coated with a thin film of metal under vacuum. As compared to shadow casting, complete coating is essential because the scanning beam produces a charge on the uncoated biological materials which cause distortion of image. The metal coated specimen is scanned with a beam of electrons (50-100 Å) which strikes on the specimen and emits the secondary electrons. These electrons pass through a cathode tube and image is produced on the screen of Cathode Ray Tube. In plant system, SEM is used to see the nature of appendages, pollen morphology and morphogenesis in plant tissue culture. It is also used to study the structures of metals, crystals and in forensic sciences.

(c) **Light Microscopy:** Bright field microscopy is absolutely indispensable tool for cell biologists. This is required for routine observations of cells, cellular differentiation and pigmentation. Fluorescence microscopy has become a powerful tool for cell biologists, particularly for the selection of fluorescing secondary metabolite rich cells. Only these two techniques are discussed here in brief. According to wave theory, light is propagated from one place to another as wave travelling in a hypothetical medium. Light waves are described in terms of their amplitude, frequency, and wavelength. Amplitude is the maximum displacement of light path from the position of equilibrium. Frequency of light is the number of complete cycles occurring in a second. The property of light waves inversely associated with the frequency is the wavelength and is defined as the distance between corresponding points on a wave or distance between two successive peaks or crests. The velocity of light is about 186300 miles or $3 \times 10^{15}$ kms per second. A good microscope has not only good magnifying power but also good resolving power to provide finer details of the object. Thus, the basic difficulty in designing a microscope is riot the magnification, but the ability of lens system to distinguish two adjacent points as distinct and separate. This ability is known as resolving power of the microscope. The resolving power of a microscope depends upon the wavelength of light and numerical aperture (N.A.). **The minimum resolvable distance between two luminous points (v) is given by the following formula:**

$$V = \frac{0.16\lambda}{N.A.} \text{ where, } \lambda = \text{wavelength}$$

Thus, shorter the wavelength of light used and lower the N.A., greater is the resolving power. The limit of resolution of a microscope is approximately equal to 0.5 /N.A., which for a light microscope is approximately 200 nanometers (nm) or about the size of many bacterial cells. The numerical aperture of a lens is dependent on the refractive index (r) = the ratio of the speed of light in a given medium to the speed of light in a vacuum) of the medium filling the space between the specimen and the front of the objective lens and on the angle of the most oblique rays of light that can enter the objective lens ($\theta$). It is given by formula [N.A. = $\eta x \sin\theta$], that is why immersion oil is placed between the object and oil immersion lens (100 X objective).

**Colorimeter:** The most commonly used method for determining the concentration of biochemical compounds is colorimetry. It uses the property of light such that when white light passes through a coloured solution, some wavelengths are absorbed more than others. Hyaline solution can be made coloured by specific reactions with suitable reagents. These reactions are generally very sensitive to determine quantities of material in the region of millimole per litre concentration. The big advantage is that complete isolation of the compound is not necessary and the constituents of a complex mixture such as blood can be determined after little treatment. The depth of colour is directly proportional to the concentration of the compound being measured, while the amount of light absorbed is proportional to the intensity of the colour and therefore, to the concentration.

**Centrifugation:** A centrifuge is an instrument which produces centrifugal force by rotating the samples around a central axis with the help of an electric motor. Centrifuges can be categorized as the clinical type (5-10,000 rpm), refrigerated high-speed centrifuges (10,000-20,000 rpm) and ultra -centrifuges (20,000 to 80,000 rpm). With increase in rpm, the friction of rotor with air produces so much of heat that they have to be run under refrigeration (so called refrigerated centrifuge) and both refrigeration and vacuum are used in ultracentrifuge, which runs at very high rpm. For these high speeds, even the rotor has to be made of special metal to withstand the great force. There are two types of rotors, angle head and swing (Fig. 25.10). In the former, the samples are kept at an angle of about 30° to the vertical axis whereas; in the latter the samples while spinning are horizontal. Simple calculations show that for the same radius, the swinging bucket method produces more gravitational force. Ultracentrifuges are of two types – analytical and preparative model.

**Analytical model:** This consists of rotors and tubes, called cells. The instrument is designed to allow the operator to follow the progress of the substances in the cells, while the process of centrifugation is in progress. By estimating sedimentation velocity during the process, the molecular weight, purity etc. can be determined.

**Preparative model:** This is used for purification of the components of macromolecules or other substances and all determinations are made at the end of centrifugation. The instrument has no monitoring device, while large centrifugal forces are set for a fixed time period. Centrifugation is the most widely used technique for separation of various metabolites and also used to separate non-miscible liquids during extraction of secondary metabolites, e.g., water (aqueous), chloroform (organic) mixture.

A wide variety of centrifuges are available, ranging in capacity and speed. During the process of centrifugation, solid particles experience a centrifugal force, which pulls them outwards, i.e., away from the center. The velocity with which a given solid particle moves through a liquid medium is related to angular velocity.

The principle of centrifugation is that, an object moving in a circular motion at an angular velocity is subjected to an outward force (F) through a radius of rotation (r) in cms, presented as $\omega^2 [F = \omega^2 r]$.

F frequently expressed in terms of gravitational force of the Earth, commonly referred to as RCF (Relative centrifugal force) or g by the following formula [RCF $= \omega 2r / 980$]. The $\omega^2$ operating speed of the centrifuge is expressed as revolutions per minute 'rpm', which can be converted to radians by the following formula:

$$\omega = \frac{rpm}{30} \frac{\pi\,(rpm)}{30}$$

or,

$$RCF = \frac{\dfrac{(\pi\,rpm)^2\,(r)}{30^2}}{980} = \frac{980\,(\pi\,rpm)^2\,(r)}{30^2}$$

Therefore,     $RCF = 1.2 \times 10^{-5}\,(rpm)^2\,r$

Sometimes the velocity of the moving particles is expressed in the form of sedimentation coefficient (s) by the formula [$V - s\,(\omega^2 r)$]. The sedimentation coefficient is a characteristic constant for a molecule or a particle and is a function of the size, shape and density. It is equivalent to the average velocity per unit of acceleration. The unit Svedberg (s) is often used with reference to centrifugation and is equivalent to a sedimentation coefficient of $10^{-13}$ s.

**Thermometer:** The thermometer is a device that measures temperature or temperature gradient using a variety of different principles; it comes from the Greek roots thermo, heat, and meter, to measure. A thermometer has two important elements: the temperature sensor (e.g., the bulb on a mercury thermometer) in which some physical change occurs with temperature, plus some means of converting this physical change into a value (e.g., the scale on a mercury thermometer). Industrial thermometers commonly use electronic means to provide a digital display or input to a computer. The Alcohol thermometer or Spirit thermometer is an alternative to the Mercury-in-glass thermometer, and functions in a similar way. An organic liquid is contained in a glass bulb which is connected to a capillary of the same glass and the end is sealed with an expansion bulb. The space above the liquid is a mixture of nitrogen and the vapour of the liquid.

For the working temperature range, the meniscus or interface between the liquid is within the capillary. With increasing temperature, the volume of liquid expands and the meniscus moves up the capillary. The position of the meniscus shows the temperature against an inscribed scale.

The liquid used can be pure ethanol or toluene or kerosene or Isoamyl acetate, depending on manufacturer and working temperature range. Since these are transparent, the liquid is made more visible by the addition of a red or blue dye. One half of the glass containing the capillary is usually enameled white or yellow to give a background for reading the scale. Temperature is measured by maximum-minimum thermometer or a continuous rotary dram chart type thermometer. The U-shaped maximum-minimum thermometer is commonly used for determining diurnal maximum and minimum range of temperature in the culture room. During the dark period, temperature remains slightly (1-2 °C) lower than light period (all bulbs and tube lights of the culture room increase the temperature). Therefore, chokes of the tube lights are fitted outside the culture room. The indicators of the maximum-minimum thermometer are moved by mercury column and they remain at that position until moved by the observer with the help of magnet. Their positions indicate the minimum and maximum temperature in the previous 24 h. After recording temperature indicators are reset to mercury level.

**Hygrometer:** Hygrometers are instruments used for measuring humidity. A simple form of a hygrometer is specifically known as a "psychrometer" and consists of two thermometers, one of which includes a dry bulb and the other of which includes a bulb that is kept wet to measure wet-bulb temperature. Evaporation from the wet bulb lowers the temperature, so that the wet-bulb thermometer usually shows a lower temperature than that of the dry-bulb thermometer, which measures dry-bulb temperature. Relative humidity is computed from the ambient temperature as shown by the dry-bulb thermometer and the difference in temperatures as shown by the wet-bulb and dry-bulb thermometers. Relative humidity can also be determined by locating the intersection of the wet- and dry-bulb temperatures on a psychrometric chart. Dial type hair hygrometer is a convenient tool to measure humidity in the culture room. Humidifier is used to maintain 60% relative humidity (RH) in the culture room. Humidistat is a bimetallic thermocouple device to control and regulate the function of humidifier to maintain the humidity. Distilled water should be filled in the humidifier. The RH present in the culture room is measured by hair hygrometer. As the name suggests a chemically treated hair elongates with increased humidity and shortens with dryness (similar to mercury in thermometer). It is calibrated from 0 to 100% RH. At lower humidity, medium dries rapidly whereas at higher humidity chances of fungal growth over all surfaces and cotton plugs is increased. Therefore, about 60% RH is maintained in the culture room.

**LUX meter:** Light is the form of radiant energy, i.e., electromagnetic radiation of specific wavelength. Visible light as we perceive, is located in narrow wavelength region of spectrum between 380 to 760 nm. Light has dual characters, displaying both wave properties (refraction, diffraction, interference and polarization phenomena) and particle properties (light is radiated in discrete amounts of energy or photons). Properties of light should be defined either as irradiance (radiant flux intercepted per unit area; unit = $w/m^2$) or an illuminance (luminous flux intercepted per unit area, unit lux, 10.76 lux = 1 foot candle). It is to note that an irradiance measurement is not spectrally defined, whereas, an illuminance measurement indicates the level of visible light as the human eye would see it.

Artificial light is provided by cool, white, fluorescent tubes and/or incandescent bulbs of different classes of artificial light sources, fluorescent sources have been used almost exclusively for plant tissue culture because they are more efficient at producing broad band visible light than incandescent bulbs and are available in low output wattage than lamps.

Light intensity can be measured by photometer or lux meter. A photometer consists of photoelectric cell and a micro-ammeter. Photoelectric cell is sensitive for light and converts light into current. Micro-ammeter shows reading due to this current and its needle moves. Nowadays digital read out is given by appropriately converting the current into digital signal (Fig. 25.12). High intensity is proportional to the current generated in the photoelectric cell by falling light. The unit of illumination is called as 'lux' and the scale of ammeter is calibrated in lux. Usually 2000 to 2500 lux is provided to the cultures maintained in light in the culture room. The instrument is a sensitive tool and handled with care. It should not be exposed to the sunlight without switching the proper reading switch to high light illumination.

## Lab Safety

1. Apparel: It is better not to step in to the laboratory wearing shoes. The footwear sold be kept outside the lab and entry should be done only after thoroughly washing the feet with sterilants. At the same time sterilized shoes can be worn at all times provided they are clean and used exclusively for PTC purpose. It is best not to wear open toed shoes or sandals, since they offer no protection from spills. We will be using substances that can ruin your clothes, e.g. bleach, so if you have a lab coat, you may want to wear it. We will not be using many dangerous chemicals, but safety glasses will be provided when we are using acids, etc.

2. No eating, drinking or smoking in the laboratory at any time.

3. Spills: If you spill a chemical on yourself, wash immediately with copious amounts of water and notify the TA or me. In the event of a spill on the floor or a bench involving hazardous materials (such as strong acid or base or a volatile organic compound) notify us immediately and receive instructions regarding clean-up before attempting to clean it up yourself.

4. Accidents: Be careful! Pay attention to what you are doing at all times. If you injure yourself in the laboratory in any way (however minor you may think the injury is), report it to the authorities immediately.

5. Broken glass: Everybody breaks glass occasionally. If you break something, do not rush to clean it up with your hands. Find a broom or dust brush, sweep up the glass and place it in the appropriate broken glassware container. Do not ever put any glass in the regular trash can.

6. Other waste: Do not put any waste chemicals down the sink. The disposal should be done into specific bins. All transgenic material must be autoclaved before disposal. Tissue culturists will also be using sharp scalpels, needles, and syringes. They must be disposed of in a special container, not in the trash.

7. Pipetting: Mouth pipetting is forbidden. Use pipettes at all times.

8. Volatile chemicals: Use the fume hood when working with volatile chemicals. Check to make sure the hood is working before opening the volatile chemical.

9. Dirty Lab ware: Follow the instructions on how to deal with dirty lab ware.

10. Labelling: Make sure that all of your cultures, etc. are properly labelled and materials are stored where instructed.

11. Transgenic procedures: Genetic manipulation experiments must be carried out in accordance with guidelines laid down by National Institutes of Health, and our local Environmental Health and Safety on our campus. We will instruct you in this.

**Sterile (Aseptic) Techniques**: Aseptic technique is absolutely necessary for the successful establishment and maintenance of plant cell, tissue and organ cultures. The in vitro environment in which the plant material is grown is also ideal for the proliferation of microorganisms. In most cases the microorganisms outgrow the plant tissues, resulting in their death. Contamination can also spread from culture to culture. The purpose of aseptic technique is minimize the possibility that microorganisms remain in or enter the cultures. The environmental control of air is also of concern because room air may be highly contaminated. Example: Sneezing produces 100,000 - 200,000 aerosol droplets which can then attach to dust particles. These contaminated particles may be present in the air for weeks. Air may also contain bacterial and fungal spores, which are likely to contaminate the cultures.

## I. Contaminants

### A. Bacteria, fungi, and insects

1. **Bacteria:** Bacteria are the most frequent contaminants. They are usually introduced with the explant and may survive surface sterilization of the explant because they are in interior tissues. So, bacterial contamination can first become apparent long after a culture has been initiated (see below). Some bacterial spores can also survive the sterilization procedure even if they are on the tissue surface. Many kinds of bacteria have been found in plant tissue cultures including *Agrobacterium, Bacillus, Corynebacterium, Enterobacter, Lactobacillus, Pseudomanas, Staphylococcus*, and *Xanthomonas*. Bacteria can be recognized by a characteristic "ooze"; the ooze can be many colors including white, cream, pink, and yellow. There is also often a distinctive odor.

2. **Fungi:** Fungi may enter cultures on explants or spores may be airborne. Fungi are frequently present as plant pathogens and in soil. They may be recognized by their "fuzzy" appearance, and occur in a multitude of colors.

3. Yeast: Yeast is a common contaminant of plant cultures. Yeasts live on the external surfaces of plants and are often present in the air.

4. **Viruses and other organisms:** Viruses, mycoplasma-like organisms, spiroplasmas, and rickettsias are extremely small organisms that are not easily detected. Thus, plant culture is not necessarily pathogen-free even if microorganisms are not detected, and this can influence culture success. Special measures such as meristem culture are often necessary to eradicate such contaminants.

5. **Insects:** The insects that are most troublesome in plant cultures include ants, thrips, and mites. Thrips often enter cultures as eggs present on the explants. Ants and mites, however, usually infest already established cultures. Mites feed on fungus and mite infestations are often first detected by observing lines of fungal infection that lead from the edge of the culture vessel to the plant tissue, having been introduced by the insect. It is very difficult to eradicate insect infestations. Careful lab practices and cleanliness should prevent most infestations.

**B. Initial contaminants:** Most contamination is introduced with the explant because of inadequate sterilization or just very dirty material. It can be fungal or bacterial. This kind of contamination can be a very difficult problem when the plant explant material is harvested from the field or greenhouse. Initial contamination is obvious within a few days after cultures are initiated. Bacteria produce "ooze" on solid medium and turbidity in liquid cultures. Fungi look "furry" on solid medium and often accumulate in little balls in liquid medium.

**C. Latent contamination:** This kind of contamination is usually bacterial and is often observed long after cultures are initiated. Apparently the bacteria are present endogenously in the initial plant material and are not obviously pathogenic in situ. Once in vitro, however, they increase in titer and overrun the cultures. Latent contamination is particularly dangerous because it can easily be transferred among cultures.

**D. Introduced contamination:** Contamination can also occur as a result of poor sterile technique or dirty lab conditions. This kind of contamination is largely preventable with proper care.

**E. Detection of contaminants:** Contamination is usually detected by the "eyeball" method in research labs. However, indexing is possible, and is frequently done in commercial settings. This involves taking a part of the plant tissue and culturing it in media that are specific for bacteria and fungi. Media that have been used for this purpose include PDA (potato dextrose agar) and NB broth (with salts, yeast extract and glucose). This is the most reliable method for detecting bacteria and fungi, but, as indicated above, there may be infecting organisms that will not be detected.

**II. Inoculation chamber:** Laminar airflow hoods are used in commercial and research tissue culture settings. A horizontal laminar flow unit is designed to remove particles from the air. Room air is pulled into the unit and pushed through a HEPA (High Energy Particle Air) filter with a uniform velocity of 90 ft/min across the work surface. The air is filtered by a HEPA filter so nothing larger than 0.3 micrometer, which includes bacterial and fungal spores, can pass through. This renders the air sterile. The positive pressure of the air flow from the unit also discourages any fungal spores or bacteria from entering. Depending on the design of the hood, the filters are located at the back or in the top of the box.

**III. Sterilization and Use of Supplies and Equipment:**

A. Sterilizing tools, media, vessels etc.

1. **Autoclaving**: Autoclaving is the method most often used for sterilizing heat-resistant items and our usual method for sterilizing items. In order to be sterilized, the item must be held at 121°C, 15 psi, for at least 15 minutes. It is important that items reach this temperature before timing begins. Therefore, time in the autoclave will vary, depending on volume in individual vessels and number of vessels in the autoclave. Most autoclaves automatically adjust time when temperature and psi are set, and include time in the cycle for a slow decrease in pressure. There are tape indicators that can be affixed to vessels, but they may not reflect the temperature of liquid within them. There are also "test kits" of microorganisms that can be run through the autoclave cycle and then cultured. Empty vessels, beakers, graduated cylinders, etc., should be closed with a cap or aluminum foil. Tools should also be wrapped in foil or paper or put in a covered sterilization tray. It is critical that the steam penetrate the items in order for sterilization to be successful.

2. **Autoclaving and Filter-sterilizing media and other liquids**: Two methods (autoclaving and membrane filtration under positive pressure) are commonly used to sterilize culture media. Culture media, distilled water, and other heat stable mixtures can be autoclaved in glass containers that are sealed with cotton plugs, aluminum foil, or plastic closures. However, solutions that contain heat-labile components must be filter-sterilized. For small volumes of liquids (100 ml or less), the time required for autoclaving is 15-20 min, but for larger quantities (2-4 liter), 30-40 min is required to complete the cycle. The pressure should not exceed 20 psi, as higher pressures may lead to the decomposition of carbohydrates and other components of a medium. Too high temperatures or too long cycles can also result in changes in properties of the medium. Organic compounds such as some growth regulators, amino acids, and vitamins may be degraded during autoclaving. These compounds require filter sterilization through a 0.22 μm membrane. Several manufacturers make nitrocellulose membranes that can be sterilized by autoclaving. They are placed between sections of a filter unit and sterilized as one piece. Other filters (the kind we use) come pre-sterilized. Larger ones can be set over a sterile flask and a vacuum is applied to pull the compound dissolved in liquid through the membrane and into the sterile flask. Smaller membranes fit on the end of a sterile syringe and liquid is pushed through by depressing the top of the syringe. The size of the filter selected depends on the volume of the solution to be sterilized and the components of the solution.

Nutrient media that contain thermo labile components are typically prepared n several steps. A solution of the heat-stable components is sterilized in the usual way by autoclaving and then cooled to 35°-50° C under sterile conditions. Solutions of the thermo labile components are filter-sterilized. The sterilized solutions are then combined under aseptic conditions to give the complete medium.

In spite of possible degradation, however, some compounds that are thought to be heat labile are generally autoclaved if results are found to be reliable and reproducible. These compounds include ABA, IAA, IBA, kinetin, pyridoxine, 2-ip and thiamine are usually autoclaved.

3. **Ethylene oxide gas:** Plastic containers that cannot be heated are sterilized commercially by ethylene oxide gas. These items are sold already sterile and cannot be resterilized. Examples of such items are plastic petri dishes, plastic centrifuge tubes etc.

4. **UV radiation:** It is possible to use germicidal lamps to sterilize items in the transfer hood when no one is working there. We do not do this. UV lamps should not be used when people are present because the light is damaging to eyes and skin. Plants left under UV lamps will die.

5. **Microwave:** It is also possible to sterilize items in the microwave; we do not do this.

6. **More comments:** Know which of your implements, flasks, etc. are sterile and which are not. Sterile things will have been autoclaved and should be wrapped with some kind of protective covering, e.g. foil, for transport from the autoclave to the hood. Our usual autoclave time of 20 minutes is intended for relatively small volumes. Large flasks of media, water, etc. may require longer autoclaving periods. It is preferable to put no more than one liter of liquid in a container to be autoclaved. Also, be sure to leave enough room in the container so that the liquid does not boil over. Sterilized items should be used within a short time (a few days at most). Items that come packaged sterile, e.g. plastic petri plates, should be examined carefully for damage before use. If part of a package is used, seal up the remainder and date and label. Use up these items unless there is some question about their sterility; they are expensive.

**IV. Working in the Transfer Hood:** The hood should remain on continuously. If for some reason it has been turned off, turn it on and let it run for at least 15 minutes before using. Make sure that everything needed for the work is in the hood and all unnecessary things are removed. As few things as possible should be stored in the hood. Check the bottom of the hood to make sure there is no paper or other debris blocking air intake. Remove watches, etc., roll up long sleeves, and wash hands thoroughly with soap (preferably bactericidal) and water. Spray or wipe the inside of the transfer hood (bottom and sides, not directly on the filters) with 70% ethyl alcohol. Others use disinfectants such as Lysol®. Wipe the work area and let the spray dry. Wipe hands and lower arms with 70% ethyl alcohol.

Spray everything going into the sterile area with 70% ethanol. For example, spray bags of petri dishes with 70 % alcohol before you open them and place the desired number of unopened dishes in the sterile area.

Work well back in the transfer hood (behind the line). Especially keep all flasks as far back to the back of the hood as possible. Movements in the hood should be contained to small areas. A line drawn across the distance behind which one should work is a useful reminder. Make sure that materials in use are to the side of your work area, so that airflow from the hood is not blocked. Do not touch any surface that is supposed to remain sterile with your hands. Use forceps, etc.

Instruments (scalpels, forceps) can be sterilized by flaming - dipping them in 95% ethyl alcohol and then immediately placing them in the flame of an alcohol lamp or gas burner. This can be dangerous if the vessel holding the alcohol tips over and an alcohol fire results. A fairly deep container, like a coplin-staining jar, should be used to hold the ethanol. Use enough ethanol to submerge the business ends of the instruments but not so much that you burn your hands. Some people wear gloves in the hood for certain procedures. If you do this, be very careful not to get them near the flame. Other methods of sterilization that do not require alcohol are with a bacticinerator or glass bead sterilizer. There is not as much risk from fire with these, but the instruments can still get extremely hot, causing burns.

Arrange tools and other items in the hood so that your hands do not have to cross over each other while working. For a right-handed person, it is best that the flame, alcohol for flaming, and tools be placed on the right. The plant material should be placed to the left. All other items in the hood should be arranged so that your work area is directly in front of you, and between 8 and 10 inches in from the front edge. No materials should be placed between the actual work area and the filter. Keep as little in the hood as possible.

Plant material should be placed on a sterile surface when manipulating it in the hood. Sterile petri dishes (expensive), sterile paper towels, or sterile paper plates work fine. Pre-sterilized plastic dishes have two sterile surfaces-the inside top and inside bottom.

Sterilize your instruments often, especially in between individual petri plates, flasks, etc. The tools should be placed on a holder in the hood to cool or should be cooled by dipping in sterile water or medium before handling plant tissues. Wipe up any spills quickly; use 70% Ethyl alcohol for cleaning. Clean hood surface periodically while working. Use of glass or plastic pipettes: Glass pipettes are put into containers or wrapped and then autoclaved. Plastic pipettes are purchased presteurilized in individual wrappers. To use a pipette, remove it from its wrapper or container by the end opposite the tip. Do not touch the lower two-thirds of the pipette. Do not allow the pipette to touch any laboratory surface. Insert only the untouched lower portion of the pipette into a sterile container.

Sterilize culture tubes with lids or caps on. When you open a sterile tube, touch only the outside of the cap, and do not set the cap on any laboratory surface. Instead, hold the cap with one or two fingers while you complete the operation, and then replace it on the tube. This technique usually requires some practice, especially if you are simultaneously opening tubes and operating a sterile pipette. After you remove the cap from the test tube, pass the mouth of the tube through a flame. If possible, hold the open tube at an angle. Put only sterile objects into the tube. Complete the operation as quickly as you reasonably can, and then flame the mouth of the tube again. Replace the lid. Inoculating loops and needles are the primary tools for transferring microbial cultures. We use plastic ones that come sterile. If you are moving organisms from an agar plate, touch an isolated colony with the transfer loop. Replace the plate lid. Open and flame the culture tube, and inoculate the medium in it by stirring the end of the transfer tool in the medium. If you are removing cells from a liquid culture, insert the loop into the culture. Even if you cannot see any liquid in the loop, there will be enough cells there to inoculate a plate or a new liquid culture.

If you do not have to be careful about the volume you transfer, a pure culture or sterile solution can be transferred to a sterile container or new sterile medium by pouring. For example, we do not measure a specific volume of medium when we pour culture plates, although after you have done it for a while, you become pretty consistent. Remove the cap or lid from the solution to be transferred. Thoroughly flame the mouth of the container, holding it at an angle as you do so. Remove the lid from the target container. Hold the container at an angle. Quickly and neatly pour the contents from the first container into the second. Replace the lid.

If you must transfer an exact volume of liquid, use a sterile pipette or a sterile graduated cylinder. When using a sterile graduated cylinder, complete the transfer as quickly as you reasonably can to minimize the time the sterile liquid is exposed to the air. Remove items from the hood as soon as they are no longer needed. All cultures must be sealed before leaving the hood. When transferring plant cultures, do contaminated cultures at the end. Situate the cultures so that the contaminated part is closest to the front of the hood.

Place waste in the proper containers: Empty (e.g. after transfer) or old petri plates used in transformation experiments go in the big bag to be autoclaved, as do other disposable that were in contact with recombinant bacterial or plant material. All needles go in the sharps box, needles used with bacteria get autoclaved. Small bags used in the hood for waste go in the big bag to be autoclaved; do not overfill the small bags or leave full bags in or on the hood for someone else to dispose of. Glassware that comes in contact with bacteria is placed in a separate pan to be autoclaved.

When finished in the hood, clean up after yourself. Remove all unnecessary materials and wipe the hood down with 70% ethyl alcohol.

Be sure when you are finished that you turn off the gas to the burner!

It is pointless to practice good sterile technique in a dirty lab. Special problems are contaminated cultures, dirty dishes and solutions where microorganisms can grow.

Store cultures in a sequestered area. We will discuss this area later. Check cultures every 3-5 days for contamination.

## V. Surface-sterilizing Plant Material

1. Preparation of stock plants: Prior good care of stock plants may lessen the amount of contamination that is present on explants. Plants grown in the field are typically more "dirty" than those grown in a greenhouse or growth chamber, particularly in humid areas like Florida. Overhead watering increases contamination of initial explants. Likewise, splashing soil on the plant during watering will increase initial contamination. Treatment of stock plants with fungicides and/ or bactericides is sometimes helpful. It is sometimes possible to harvest shoots and force buds from them in clean conditions. The forced shoots may then be free of contaminants when surface-sterilized in a normal manner. Seeds may be sterilized and germinated in vitro to provide clean material. Covering growing shoots for several days or weeks prior to harvesting tissue for culture may supply cleaner material. Explants or material from which material will be cut can be washed in soapy water and then placed under running water for 1 to 2 hours.

2. Ethanol (or Isopropyl Alcohol): Ethanol is a powerful sterilizing agent but also extremely phytotoxic. Therefore, plant material is typically exposed to it for only

seconds or minutes. The more tender the tissue, the more it will be damaged by alcohol. Tissues such as dormant buds, seeds, or unopened flower buds can be treated for longer periods of time since the tissue that will be explanted or that will develop is actually within the structure that is being surface-sterilized. Generally, 70% ethanol is used prior to treatment with other compounds.

3.  Sodium hypochlorite: Sodium hypochlorite, usually purchased as laundry bleach, is the most frequent choice for surface sterilization. It is readily available and can be diluted to proper concentrations. Commercial laundry bleach is 5.25% sodium hypochlorite. It is usually diluted to 10% - 20% of the original concentration, resulting in a final concentration of 0.5 - 1.0% sodium hypochlorite. Plant material is usually immersed in this solution for 10 - 20 minutes. A balance between concentration and time must be determined empirically for each type of explant, because of phytotoxicity.

4.  Calcium hypochlorite: Calcium hypochlorite is used more in Europe than in the U.S. It is obtained as a powder and must be dissolved in water. The concentration that is generally used is 3.25 %. The solution must be filtered prior to use since not all of the compound goes into solution. Calcium hypochlorite may be less injurious to plant tissues than sodium hypochlorite.

5.  Mercuric Chloride: Mercuric chloride is used only as a last resort in the U.S. It is extremely toxic to both plants and humans and must be disposed of with care. Since mercury is so phytotoxic, it is critical that many rinses be used to remove all traces of the mineral from the plant material.

6.  Hydrogen peroxide: The concentration of hydrogen peroxide used for surface sterilization of plant material is 30%, ten times stronger than that obtained in a pharmacy. Some researchers have found that hydrogen peroxide is useful for surface-sterilizing material while in the field.

7.  Enhancing Effectiveness of Sterilization Procedure

    »   Surfactant (e.g. Teepol) is frequently added to the sodium hypochlorite.

    »   A mild vacuum may be used during the procedure.

    »   The solutions that the explants are in are often shaken or continuously stirred.

8.  **Rinsing**: After plant material is sterilized with one of the above compounds, it must be rinsed thoroughly with sterile water. Typically, three to four separate rinses are done.

9. **Use of antibiotics and fungicides in vitro**: We have found that the use of antibiotics and fungicides in vitro is not very effective in eliminating microorganisms and these compounds are often quite phytotoxic.

10. **Plant preservative mixture:** PPM™ is a proprietary broad-spectrum biocide, which can be used to control contamination in plant cell cultures, either during the sterilization procedure, or as a medium component. PPM™ comes in an acidic liquid solution (pH 3.8). The recommended dose is 0.5–2.0 mL of PPM™ per liter of medium. Higher doses are required to treat endogenous contamination and for Agrobacterium. Its makers say that PPM™ has several advantages over antibiotics: It is effective against fungi as well as bacteria, thus it can be substituted for a cocktail of antibiotics and fungicides. PPM™ is less expensive than antibiotics, which makes it affordable for wide and routine use. The formation of resistant mutants toward PPM™ is very unlikely because it targets and inhibits multiple enzymes. Many antibiotics adversely affect plant materials. If used as recommended, PPM™ does not adversely affect in vitro seed germination, callous proliferation, or callous regeneration. Seeds and explants with endogenous contamination can be sterilized at doses of 5-20 mL/L of PPM™. This is useful when routine surface sterilization is insufficient.

Basic Laboratory Procedures Involved in Media Making: The majority of laboratory operations utilized in the in vitro propagation of plants can be easily learned. One needs to concentrate mainly on accuracy, cleanliness, and strict adherence to details when performing in vitro techniques.

1. **Medium Stock Solutions:** In the old days (when I was in graduate school), mineral salt mixtures were prepared as stock solutions ranging from 10 to 100 times the final concentrations used in the medium. Stock solutions can be prepared as two solutions, one containing all of the macronutrients and one containing all of the micronutrients. These solutions must be kept fairly dilute (10-20X) in order to avoid precipitation of calcium and magnesium phosphates and sulphates. A more common method is to arrange mineral salt stock solutions according to the ions they contain. A series of solutions containing the inorganic components of the medium is prepared; precise combinations may vary from lab to lab. Using this method, salt stocks can be prepared in 100X concentrations. Iron-EDTA chelate is prepared from iron sulphate and Na-EDTA by mixing the proper amounts of the two compounds with water and then autoclaving. This stock must then be stored in a dark container. The prepared stock solutions are usually stored in the refrigerator. Although the initial cost of chemicals may be substantial, overall the ongoing cost of using stock solutions is probably less than that of using prepared mixes (unless

you factor in labor). However, the stocks must be prepared over and over when they are used up, and each time this is done potential error is introduced. It is best if one person consistently prepares the stocks. Most people no longer make up stock solutions of the inorganic medium components. Stocks solutions may be useful if many different media are made in the lab or frequent changes in concentrations are made in individual components.

**II. Prepared Mixes:** Several companies sell prepared salt and vitamin mixtures as powders. These are easily handled by adding the proper amount of powder to water. The mixtures can be purchased as complete media or as salts alone. The packs often contain the necessary ingredients for one liter of medium. The contents are extremely hygroscopic once a pack is opened, so it is best to use the contents all at once. A small amount of precipitate may be seen in media prepared using either of these methods. This results from iron being displaced from the chelate over time as media is stored. It doesn't appear to be deleterious.

**III. Organic Addenda:** Most organic addenda are added in relatively low concentrations, too low to be weighed out accurately. Therefore, stock solutions ranging from 100 to 1000X final concentrations are prepared. Vitamins may be prepared as stocks or purchased premade. Vitamin solutions can be divided into aliquots and stored in the freezer at -20° C. Most growth regulators are stable for up to a month in stock solutions. The solvents used for dissolving growth regulators vary depending on the compound under consideration.

IV. Making Stock Solutions: There are three ways to describe amounts of chemicals in a stock: parts per million (rarely used), as grams per liter, or as molar units (journals usually require concentrations to be expressed this way).

**Major equipment and their functions**

| S. No. | Name of the equipment | Functions |
|--------|----------------------|-----------|
| 1. | Autoclave machine, Pressure cooker | Sterilization of media, glassware &small instrument. |
| 2. | Balance | Measurement of chemical from the range of µgm to Kg |
| 3. | Hot plate magnetic stirrer | To mix the chemical & other ingredient of media |
| 4. | pH meter | To determine the pH of various chemicals & media |

| 5. | Refrigerator | To store all sorts of temperature-sensitive chemical & stock solution. |
|---|---|---|
| 6. | Micro oven | To melt agar, agarose & other gelling agents. |
| 7. | Hot air oven | For dry heat sterilization of cell & suspension culture |
| 8. | Shaker | Use for gentle rotation of cell& suspension culture |
| 9. | Filter sterilization unit with vacuum pump | Filtration of thermoliable compound like growth regulator, vitamin, amino acid etc. |
| 10. | Microscope | To study the cell & tissue culture material at different stages of development |
| 11. | Luxmeter | To measure the light intensity of the culture room |
| 12. | Thermometer | To record the temperature reading of laboratory & culture room |
| 13. | Centrifuge machine | To sediment cell & clean supernatant |

# Chapter - 6

## Basic Techniques of Plant Tissue Culture

### Selection of the Plant Material

Selection of the Plant material is the first and foremost step in PTC. Objectives of the research project actually determines this step. Generally, useful, economically valuable, commercially significant or rare & endangered plants are selected for the purpose of PTC. The most advantageous characteristic of the tissue culture is the ability to utilize "any piece of tissue from any part of the plant" to grow another plant. Other advantages include the production of a large number of plants in a shorter timespan compared to conventional techniques, production of disease-free plants, etc. The piece of plant tissue that's placed on the growth media is called the explant. Though it's known that any plant tissue can be used to grow the entire plant, there are some conditions applied! There's only type of plant tissue that's not suitable to grow all species of plants in tissue culture. Some factors determine which explant will be suitable for your culture; by carefully choosing healthy explants for your culture, you can increase the culture productivity multifold.

### Types of Explants Used in Tissue Culture

All parts of plants can be used as explants for tissue culture purposes: leaves, stems, a portion of shoots, flowers, anthers, ovary, single undifferentiated cells, mature tissues single cells or protoplasts. Basically, any living cell or part of the plant can be used as the source of plant tissue culture. However, the cells should be capable of de-differentiating and resume cell division and have cytoplasm. Some parts, like the root tip, are rarely used as explants for tissue culture processes. Root tips are difficult to isolate and contain microbes in their tissues, forming a strong association with them. Also, soil particles are attached to roots which are difficult to remove without damaging some tissues.

## Factors to be Considered while Choosing Explants

1. **Age of the Organ used to Source the Explant:** The age of the explant is an important factor when choosing the right explant. It's advised by researchers to use young parts as a source of tissue for culturing because younger tissues correlate to better physiological responses in laboratory research. Since young tissues are newly generated, they are better suited for the rough process of surface sterilization. Surface sterilization is a process that can potentially damage tissue and helps to establish clean cultures.

2. **Season of Explant collection:** The season in which the explant is collected impacts its contamination and growth response in culture. Different seasons have different effects on the explants. For example, the explants (like buds or shoots) collected during the spring (while in their flush state) are more responsive compared to explants collected during the winter season (while in their dormant state). The dormant explants are generally unresponsive in culture, meaning they must undergo several chemical treatments to make them responsive to tissue culturing processes. Moreover, the chances of contamination also increase toward the winter time.

3. **Size and Place of the Explant:** Smaller explants are difficult to culture and are less responsive than the larger size explants. Larger pieces of tissues contain enough nutrient reserves and plant growth regulators to stimulate their growth in cultures. Smaller explants require additional components to sustain the culture. The other factor that decides the health of cultures is source of extraction from the plant. For example, a report published by Tran Thah Van (1977) shows that depending on whether the explants are taken from the base, middle, or top of the stem, the growth regulators in explants also vary. The growth hormone level of explants shows different in vitro responses.

4. **Quality of the Source Plant:** The health of the plant is an essential factor to be considered before obtaining the explant. Researchers advise taking explants from the healthy plants compared to plant under nutritional or water stress or showing any disease symptoms.

5. **Purpose of Tissue Culture:** Before choosing the explant for your culture, you should consider what value you want to receive from the culture. Depending on the desired response, the choice of explants varies. For example, for

clonal propagation, the suitable explant will be a lateral or terminal bud shoot or shoot; for callus induction, pieces of the cotyledon, hypocotyl, stem, leaf, or embryo are usually used. For haploid production, anther or pollen is cultured.

6. **Plant Genotype:** All species of the plants don't equally respond in tissue culture. Large differences exist between different genotypes, species, or cultivars, making some easily grown and some non-responsive or recalcitrant.

## Basic Prerequisites of PTC Laboratory

1. **Apparel:** Shoes are to be worn at all times. It is best not to wear open toed shoes or sandals, since they offer no protection from spills. We will be using substances that can ruin your clothes, e.g., bleach, so if you have a lab coat, you may want to wear it. We will not be using many dangerous chemicals, but safety glasses will be provided when we are using acids, etc.

2. No eating, drinking or smoking in the laboratory at any time.

3. Spills: If you spill a chemical on yourself, wash immediately with copious amounts of water and notify the TA or me. In the event of a spill on the floor or a bench involving hazardous materials (such as strong acid or base or a volatile organic compound) notify us immediately and receive instructions regarding clean-up before attempting to clean it up yourself.

4. Accidents: Be careful! Pay attention to what you are doing at all times. If you injure yourself in the laboratory in any way (however minor you may think the injury is), report it to us immediately.

5. Broken Glass: Everybody breaks glass occasionally. If you break something, don't rush to clean it up with your hands. Find a broom or dust brush, sweep up the glass and place it in the appropriate broken glassware container. Do not ever put any glass in the regular trash can.

6. Other Waste: Do not put any waste chemicals down the sink. We will instruct you as to disposal. All transgenic material must be autoclaved before disposal. We will also be using sharps, e.g. needles, and other "hospital" supplies, e.g. syringes. These must be disposed of in a special container, not in the trash.

7. Pipetting: Mouth pipetting is forbidden. Use pipettors at all times.

8. Volatile Chemicals: Use the fume hood when working with volatile chemicals. Check to make sure the hood is working before opening the volatile chemical

9. Dirty Labware: Follow the TAs instructions on how to deal with dirty labware.

10. Labeling: Make sure that all of your cultures, etc. are properly labeled and materials are stored where instructed.

11. Transgenic procedures: Genetic manipulation experiments must be carried out in accordance with guidelines laid down by National Institutes of Health, and our local Environmental Health and Safety on our campus. We will instruct you in this.

## Sterile (Aseptic) Technique

Aseptic technique is absolutely necessary for the successful establishment and maintenance of plant cell, tissue and organ cultures. The in vitro environment in which the plant material is grown is also ideal for the proliferation of microorganisms. In most cases the microorganisms outgrow the plant tissues, resulting in their death. Contamination can also spread from culture to culture. The purpose of aseptic technique is minimize the possibility that microorganisms remain in or enter the cultures.

The environmental control of air is also of concern because room air may be highly contaminated. Example: Sneezing produces 100,000 - 200,000 aerosol droplets which can then attach to dust particles. These contaminated particles may be present in the air for weeks. (Have you ever viewed the air around you when you open the curtains on a sunny day?)...Air may also contain bacterial and fungal spores, as do we.

## I. Contaminants

A. Bacteria, fungi, and insects

1. Bacteria: Bacteria are the most frequent contaminants. They are usually introduced with the explant and may survive surface sterilization of the explant because they are in interior tissues. So, bacterial contamination can first become apparent long after a culture has been initiated (see below). Some bacterial spores can also survive the sterilization procedure even if they are on the tissue surface. Many kinds of bacteria have been found in plant tissue cultures including Agrobacterium, Bacillus, Corynebacterium, Enterobacter, Lactobacillus, Pseudomanas, Staphylococcus, and Xanthomonas. Bacteria can be recognized by a characteristic "ooze"; the ooze can be many colors including white, cream, pink, and yellow. There is also often a distinctive odor.

2. Fungi: Fungi may enter cultures on explants or spores may be airborne. Fungi are frequently present as plant pathogens and in soil. They may be recognized by their "fuzzy" appearance, and occur in a multitude of colors.

3. Yeast: Yeast is a common contaminant of plant cultures. Yeasts live on the external surfaces of plants and are often present in the air.

4. Viruses, etc.: Viruses, mycoplasma-like organisms, spiroplasmas, and rickettsias are extremely small organisms that are not easily detected. Thus, plant culture is not necessarily pathogen-free even if microorganisms are not detected, and this can influence culture success. Special measures such as meristem culture are often necessary to eradicate such contaminants.

5. Insects: The insects that are most troublesome in plant cultures include ants, thrips, and mites. Thrips often enter cultures as eggs present on the explants. Ants and mites, however, usually infest already established cultures. Mites feed on fungus and mite infestations are often first detected by observing lines of fungal infection that lead from the edge of the culture vessel to the plant tissue, having been introduced by the insect. It is very difficult to eradicate insect infestations. Careful lab practices and cleanliness should prevent most infestations.

## B. Initial contaminants

Most contamination is introduced with the explant because of inadequate sterilization or just very dirty material. It can be fungal or bacterial. This kind of contamination can be a very difficult problem when the plant explant material is harvested from the field or greenhouse. Initial contamination is obvious within a few days after cultures are initiated. Bacteria produce "ooze" on solid medium and turbidity in liquid cultures. Fungi look "furry" on solid medium and often accumulate in little balls in liquid medium.

## C. Latent contamination

This kind of contamination is usually bacterial and is often observed long after cultures are initiated. Apparently the bacteria are present endogenously in the initial plant material and are not obviously pathogenic in situ. Once in vitro, however they increase in titer and overrun the cultures. Latent contamination is particularly dangerous because it can easily be transferred among cultures.

## D. Introduced contamination

Contamination can also occur as a result of poor sterile technique or dirty lab conditions. This kind of contamination is largely preventable with proper care.

## E. Detection of contaminants

Contamination is usually detected by the "eyeball" method in research labs. However, indexing is possible, and is frequently done in commercial settings. This involves taking a part of the plant tissue and culturing it in media that are specific for bacteria and fungi. Media that have been used for this purpose include PDA (potato dextrose agar) and NB broth (with salts, yeast extract and glucose). This is the most reliable method for detecting bacteria and fungi, but, as indicated above, there may be infecting organisms that won't be detected.

## II. The Transfer Hood

Laminar airflow hoods are used in commercial and research tissue culture settings. A horizontal laminar flow unit is designed to remove particles from the air. Room air is pulled into the unit and pushed through a HEPA (High Energy Particle Air) filter with a uniform velocity of 90 ft/min across the work surface. The air is filtered by a HEPA (high efficiency particulate air) filter so nothing larger than 0.3 micrometer, which includes bacterial and fungal spores, can pass through. This renders the air sterile. The positive pressure of the air flow from the unit also discourages any fungal spores or bacteria from entering. Depending on the design of the hood, the filters are located at the back or in the top of the box.

## III. Sterilization and Use of Supplies and Equipment:

Contamination is the most common problem in Plant Tissue Culture Laboratory. It is the accidental introduction of contaminants like bacteria, fungi, algae, or other sources of pathogens into the culture medium. Whether you like it or not the pathogens enter into the media either through explants or the researchers. It seriously affects the yield and productivity of cultured plants. Many techniques are available to avoid contamination from your cultures and there are mainly five methods of sterilization.

The further details of all the equipment used the process of decontamination and also the concentrations of chemicals are discussed in separate chapters.

1. **Heat or Dry sterilization:** Heat is used to sterilize the equipment and also some types of explants such as fruits. Autoclave is the most commonly

used equipment and it works on the principle of moist heat and they can sterilize glass apparatus, solid or liquid media, distilled water, normal saline, discarded cultures, and contaminated media. Hot air ovens are also used and they work on the principle of dry heat. They are used to sterilize glassware (like Petri dishes, flasks, pipettes, and test tubes), Powder (like starch, zinc oxide, and sulfadiazine), Materials that contain oils, Metal equipment (like scalpels, scissors, and blades). All the glassware and instruments are heat sterilized in the pre-heated oven at 200 °C for 1hr.

2. **Sterilization by filtration:** A liquid solution is forced through a membrane to sterilize it for microorganisms. The size of the pore in the filter decides what size of microorganisms will be filtered out. It is used to sterilize growth regulators that are thermolabile such as Thidiazuron (TDZ) NAA (Naphthalene Acetic Acid), Zeatin, Gibberellic acid (GA3), Abscisic acid (ABA), urea, and certain vitamins.

3. **Air sterilization:** Laminar flow hood is the best example. It is equipped with HEPA (High Efficiency Particulate Air) filters and a blower that blows the decontaminated clean air through the filters to prevent all sorts of microorganisms larger than 0.3 micrometers with 99% efficiency.

4. **Tyndallisation:** It is used to kill heat-resistant endospore. In this technique, the medium is heated in a water bath for 1 hour for 3 consecutive days and then kept at room temperature after two incidents of successive boiling.

5. **Surface sterilization:** This technique is used to sterilize the surface of the explant. This is the most important step in Plant tissue culture **(Details are given below in this chapter)**

**Surface sterilization:** It is an essential step in tissue culture. The explants that you obtain from the source plant should be completely free from any type of microorganisms before culturing it on the media. Some endophytes are also present in plants, microbes that live inside the explant. For endophyte prevention, you can examine the source plants and explant before using them for the culturing process. In the absence of an equipment to perform the process, you can use the Plant Preservative Mixture (PPM) in your culture media. PPM is an effective chemical that is used in tissue culture to avoid all kinds of contaminants from cultures.

**Explant sterilization is done by using the following five important chemicals.**

1. **Mercuric chloride:** It is rarely used in labs because of its high toxicity to

plants as well as humans. So, if you are using it, follow extensive care. After exposing the explants to mercuric chloride, rinse them many times using sterile water to remove all traces of mineral from the explants.

2. **Sodium hypochlorite:** It is commonly known as bleach. It is the most frequent choice for surface sterilization of explants. Commercial laundry bleach is 5.25% sodium hypochlorite. When used in tissue culture, bleach is diluted to 10-20% that results in 0.5-1.0% of bleach in the final concentration. The duration of time required to sterilize the explant varies for each type of explant. But, usually 10-20 minutes are enough to serve the purpose of sterilization.

3. **Calcium hypochlorite:** It is commercially available in the form of powder and before using it for sterilization, it must be dissolved in water. After dissolving it in the water, filter the solution. The final concentration of the solution used in labs is around 3.25%.

4. **Hydrogen Peroxide:** It is a rarely used chemical for surface sterilization. The favorable concentration of using hydrogen peroxide is 30%. Take all precautions if you are using the chemical for the surface sterilization of your explants.

5. **Ethanol:** It is also called isopropyl alcohol. The 70% ethanol is extensively used as a sterilizing agent in tissue culture labs, but It is also extremely phytotoxic. Explants are only kept for one second in the chemical because longer exposure to ethanol damages the tender explants. Methanol is also used in the same concentration.

In addition to these chemicals, in some cases, bromine, silver nitrate, and a mix of antibiotics are also used for surface sterilization of plant materials. The list of surface sterilants and the concentrations are given in the table-1

**Orchid Seed Sterilization:** Orchid seeds are very small and contain little to no food reserves. A single seed capsule may contain 1,500 to 3,000,000 seeds. Sowing the seed in vitro makes it possible to germinate immature seed (green pods). It is much easier to sterilize green capsules than individual seeds after the capsule have split open. Lucke (1971) indicated that orchid seed can be sterilized when the capsule is about two-thirds ripe. Listed below are the estimated normal ripening times of capsules for various orchid species. Soak the capsule in a 100% bleach solution for 30 minutes. Dip the capsule into 95% alcohol, and flame. Under aseptic conditions, open the capsule and scrape out the seed. Carefully layer the seed over the surface of the culture medium.

**Dry Seed Sterilization:** Collect seed and place in either a small flask or bottle, or place in a shortened pipet which has one end sealed with cotton. Seal the other end of the pipet with cotton, once the seed has been placed in the pipette. Prepare a solution containing 5-10% commercial bleach containing a few drops of TWEEN® 20. Add the bleach solution to the flask, or draw up the solution into the pipet. Swirl the flask containing the seed and bleach or repeatedly draw and aspirate the bleach solution in and out of the pipette. Sterilize the seed for 5-10 minutes. Remove the bleach solution and rinse the seed with sterile tissue culture grade water. Transfer the seed to sterile culture medium.

**Media sterilization:** Plant tissue culture media are generally sterilized by autoclaving at 121 °C and 1.05 kg/cm² (15-20 psi). The time required for sterilization depends upon the volume of medium in the vessel. The minimum times required for sterilization of different volumes of medium are listed below. It is advisable to dispense medium in small aliquots whenever possible as many media components are broken down on prolonged exposure to heat. There is evidence that medium exposed to temperatures in excess of 121 °C may not properly gel or may result in poor cell growth.

**Autoclaving the Medium:** Minimum autoclaving time includes the time required for the liquid volume to reach the sterilizing temperature (121 °C) and 15 min. at 121 °C (Burger, 1988). Times may vary due to differences in autoclaves. Validation with your system is recommended. Several medium components are considered thermolabile and should not be autoclaved. Stock solutions of the heat labile components are prepared and filter sterilized through a 0.22 μm filter into a sterile container. The filtered solution is aseptically added to the culture medium, which has been autoclaved and allowed to cool to approximately 35-45 °C. The medium is then dispensed under sterile conditions. Experimentation with your system is recommended.

**A. Sterilizing tools, media, vessels etc.**

**1. Autoclaving:** Autoclaving is the method most often used for sterilizing heat-resistant items and our usual method for sterilizing items. In order to be sterilized the item must be held at 121°C, 15 psi, for at least 15 minutes. It is important that items reach this temperature before timing begins. Therefore, time in the autoclave will vary, depending on volume in individual vessels and number of vessels in the autoclave. Most autoclaves automatically adjust time when temperature and psi are set, and include time in the cycle for a slow decrease in pressure. There are tape indicators that can be affixed to vessels, but they may not reflect the temperature of liquid within them. There are also "test kits" of microorganisms that can be run through the autoclave cycle and then cultured.

Empty vessels, beakers, graduated cylinders, etc., should be closed with a cap or aluminum foil. Tools should also be wrapped in foil or paper or put in a covered sterilization tray. It is critical that the steam penetrate the items in order for sterilization to be successful.

**2. Autoclaving and filter-sterilizing media and other liquids:** Two methods (autoclaving and membrane filtration under positive pressure) are commonly used to sterilize culture media. Culture media, distilled water, and other heat stable mixtures can be autoclaved in glass containers that are sealed with cotton plugs, aluminum foil, or plastic closures. However, solutions that contain heat-labile components must be filter-sterilized. For small volumes of liquids (100 ml or less), the time required for autoclaving is 15-20 min, but for larger quantities (2-4 liter), 30-40 min is required to complete the cycle. The pressure should not exceed 20 psi, as higher pressures may lead to the decomposition of carbohydrates and other components of a medium. Too high temperatures or too long cycles can also result in changes in properties of the medium.

Organic compounds such as some growth regulators, amino acids, and vitamins may be degraded during autoclaving. These compounds require filter sterilization through a 0.22 μm membrane. Several manufacturers make nitrocellulose membranes that can be sterilized by autoclaving. They are placed between sections of a filter unit and sterilized as one piece. Other filters (the kind we use) come pre-sterilized. Larger ones can be set over a sterile flask and a vacuum is applied to pull the compound dissolved in liquid through the membrane and into the sterile flask. Smaller membranes fit on the end of a sterile syringe and liquid is pushed through by depressing the top of the syringe. The size of the filter selected depends on the volume of the solution to be sterilized and the components of the solution.

Nutrient media that contain thermo labile components are typically prepared in several steps. A solution of the heat-stable components is sterilized in the usual way by autoclaving and then cooled to 35°-50° C under sterile conditions. Solutions of the thermo labile components are filter-sterilized. The sterilized solutions are then combined under aseptic conditions to give the complete medium.

In spite of possible degradation, however, some compounds that are thought to be heat labile are generally autoclaved if results are found to be reliable and reproducible. These compounds include ABA, IAA, IBA, kinetin, pyridoxine, 2-ip and thiamine are usually autoclaved.

**3. Ethylene oxide gas:** Plastic containers that cannot be heated are sterilized commercially by ethylene oxide gas. These items are sold already sterile and cannot be resterilized. Examples of such items are plastic petri dishes, plastic centrifuge tubes etc.

**4. UV radiation:** It is possible to use germicidal lamps to sterilize items in the transfer hood when no one is working there. We do not do this. UV lamps should not be used when people are present because the light is damaging to eyes and skin. Plants left under UV lamps will die.

**5. Microwave:** It is also possible to sterilize items in the microwave; we do not do this.

**6. More comments**

» Know which of your implements, flasks, etc. are sterile and which are not. Sterile things will have been autoclaved and should be wrapped with some kind of protective covering, e.g. foil, for transport from the autoclave to the hood.

» Our usual autoclave time of 20 minutes is intended for relatively small volumes. Large flasks of media, water, etc. may require longer autoclaving periods. It is preferable to put no more than one liter of liquid in a container to be autoclaved. Also, be sure to leave enough room in the container so that the liquid does not boil over.

» Sterilized items should be used within a short time (a few days at most).

» Items that come packaged sterile, e.g. plastic petri plates, should be examined carefully for damage before use. If part of a package is used, seal up the remainder and date and label. Use up these items unless there is some question about their sterility; they are expensive.

**IV. Working in the Transfer Hood:**

» The hood should remain on continuously. If for some reason it has been turned off, turn it on and let it run for at least 15 minutes before using.

» Make sure that everything needed for the work is in the hood and all unnecessary things are removed. As few things as possible should be stored in the hood.

» Check the bottom of the hood to make sure there is no paper or other debris blocking air intake.

» Remove watches, etc., roll up long sleeves, and wash hands thoroughly with soap (preferably bactericidal) and water.

» Spray or wipe the inside of the transfer hood (bottom and sides, not directly on the filters) with 70% EtOH. Others use disinfectants such as Lysol®. Wipe the work area and let the spray dry.

» Wipe hands and lower arms with 70% EtOH. It is not necessary to flame them (This is a joke.).

» Spray everything going into the sterile area with 70% ethanol. For example, spray bags of petri dishes with 70 % alcohol before you open them and place the desired number of unopened dishes in the sterile area.

» Work well back in the transfer hood (behind the line). Especially keep all flasks as far back to the back of the hood as possible. Movements in the hood should be contained to small areas. A line drawn across the distance behind which one should work is a useful reminder.

» Make sure that materials in use are to the side of your work area, so that airflow from the hood is not blocked.

» Don't touch any surface that is supposed to remain sterile with your hands. Use forceps, etc.

» Instruments (scalpels, forceps) can be sterilized by flaming - dipping them in 95% EtOH and then immediately placing them in the flame of an alcohol lamp or gas burner. This can be dangerous if the vessel holding the alcohol tips over and an alcohol fire results. A fairly deep container, like a coplin-staining jar, should be used to hold the ethanol. Use enough ethanol to submerge the business ends of the instruments but not so much that you burn your hands. Some people wear gloves in the hood for certain procedures. If you do this, be very careful not to get them near the flame. Other methods of sterilization that do not require alcohol are with a bacticinerator or glass bead sterilizer. There is not as much risk from fire with these, but the instruments can still get extremely hot, causing burns.

» Arrange tools and other items in the hood so that your hands do not have to cross over each other while working. For a right-handed person, it is best that the flame, alcohol for flaming, and tools be placed on the right. The plant material should be placed to the left. All other items in the hood should be arranged so that your work area is directly in front of you, and between 8 and 10 inches in from the front edge. No materials should be placed between the actual work area and the filter. Keep as little in the hood as possible.

» Plant material should be placed on a sterile surface when manipulating it in the hood. Sterile petri dishes (expensive), sterile paper towels, or sterile paper plates work fine. Pre-sterilized plastic dishes have two sterile surfaces-the inside top and inside bottom.

» Sterilize your instruments often, especially in between individual petri plates, flasks, etc. The tools should be placed on a holder in the hood to cool or should be cooled by dipping in sterile water or medium before handling plant tissues.

» Wipe up any spills quickly; use 70% EtOH for cleaning. Clean hood surface periodically while working.

» Use of glass or plastic pipettes: Glass pipettes are put into containers or wrapped and then autoclaved. Plastic pipettes are purchased presterilized in individual wrappers. To use a pipette, remove it from its wrapper or container by the end opposite the tip. Do not touch the lower two-thirds of the pipette. Do not allow the pipette to touch any laboratory surface. Insert only the untouched lower portion of the pipette into a sterile container.

» Sterilize culture tubes with lids or caps on. When you open a sterile tube touch only the outside of the cap, and do not set the cap on any laboratory surface. Instead, hold the cap with one or two fingers while you complete the operation, and then replace it on the tube. This technique usually requires some practice, especially if you are simultaneously opening tubes and operating a sterile pipette. After you remove the cap from the test tube pass the mouth of the tube through a flame. If possible, hold the open tube at an angle. Put only sterile objects into the tube. Complete the operation as quickly as you reasonably can, and then flame the mouth of the tube again. Replace the lid.

» Inoculating loops and needles are the primary tools for transferring microbial cultures. We use plastic ones that come sterile. If you are moving organisms from an agar plate, touch an isolated colony with the transfer loop. Replace the plate lid. Open and flame the culture tube, and inoculate the medium in it by stirring the end of the transfer tool in the medium. If you are removing cells from a liquid culture, insert the loop into the culture. Ever if you cannot see any liquid in the loop, there will be enough cells there to inoculate a plate or a new liquid culture.

» If you don't have to be careful about the volume you transfer, a pure culture or sterile solution can be transferred to a sterile container or new sterile medium by pouring. For example, we do not measure a specific volume of medium when we pour culture plates, although after you have done it for a while, you become pretty consistent. Remove the cap or lid from the solution to be transferred. Thoroughly flame the mouth of the container, holding it at an angle as you do so. Remove the lid from the target container. Hold the container at an angle. Quickly and neatly pour the contents from the first container into the second. Replace the lid.

» If you must transfer an exact volume of liquid, use a sterile pipette or a sterile graduated cylinder. When using a sterile graduated cylinder, complete the transfer as quickly as you reasonably can to minimize the time the sterile liquid is exposed to the air.

» Remove items from the hood as soon as they are no longer needed. All cultures must be sealed before leaving the hood.

» When transferring plant cultures, do contaminated cultures last. Situate the cultures so that the contaminated part is closest to the front of the hood.

» Place waste in the proper containers: Empty (e.g. after transfer) or old petri plates used in transformation experiments go in the big bag to be autoclaved, as do other disposable that were in contact with recombinant bacterial or plant material. All needles go in the sharps box, needles used with bacteria get autoclaved. Small bags used in the hood for waste go in the big bag to be autoclaved; do not overfill the small bags or leave full bags in or on the hood for someone else to dispose of. Glassware that comes in contact with bacteria is placed in a separate pan to be autoclaved.

» When finished in the hood, clean up after yourself. Remove all unnecessary materials and wipe the hood down with 70% EtOH.

» Be sure when you are finished that you turn off the gas to the burner!

» It is pointless to practice good sterile technique in a dirty lab. Special problems are contaminated cultures, dirty dishes and solutions where microorganisms can grow.

» Store cultures in a sequestered area. We will discuss this area later. Check cultures every 3-5 days for contamination.

## V. Surface-sterilizing Plant Material

### 1. Preparation of Stock Plants

Prior good care of stock plants may lessen the amount of contamination that is present on explants. Plants grown in the field are typically more "dirty" than those grown in a greenhouse or growth chamber, particularly in humid areas like Florida. Overhead watering increases contamination of initial explants. Likewise, splashing soil on the plant during watering will increase initial contamination. Treatment of stock plants with fungicides and/or bacteriocides is sometimes helpful. It is sometimes possible to harvest shoots and force buds from them in clean conditions. The forced shoots may then be free of contaminants when surface-sterilized in a normal manner. Seeds may be sterilized and germinated in vitro to provide clean material. Covering growing shoots for several days or weeks prior to harvesting tissue for culture may supply cleaner material. Explants or material from which material will be cut can be washed in soapy water and then placed under running water for 1 to 2 hours.

### 2. Ethanol (or Isopropyl Alcohol)

Ethanol is a powerful sterilizing agent but also extremely phytotoxic. Therefore plant material is typically exposed to it for only seconds or minutes. The more tender the tissue, the more it will be damaged by alcohol. Tissues such as dormant buds, seeds, or unopened flower buds can be treated for longer periods of time since the tissue that will be explanted or that will develop is actually within the structure that is being surface-sterilized. Generally 70% ethanol is used prior to treatment with other compounds.

### 3. Sodium Hypochlorite

Sodium hypochlorite, usually purchased as laundry bleach, is the most frequent choice for surface sterilization. It is readily available and can be diluted to proper concentrations. Commercial laundry bleach is 5.25% sodium hypochlorite. It is usually diluted to 10% - 20% of the original concentration, resulting in a final concentration of 0.5 - 1.0% sodium hypchlorite. Plant material is usually immersed in this solution for 10 - 20 minutes. A balance between concentration and time must be determined empirically for each type of explant, because of phytotoxicity.

## 4. Calcium Hypochlorite

Calcium hypochlorite is used more in Europe than in the U.S. It is obtained as a powder and must be dissolved in water. The concentration that is generally used is 3.25 %. The solution must be filtered prior to use since not all of the compound goes into solution. Calcium hypochlorite may be less injurious to plant tissues than sodium hypochlorite.

## 5. Mercuric Chloride

Mercuric chloride is used only as a last resort in the U.S. It is extremely toxic to both plants and humans and must be disposed of with care. Since mercury is so phytotoxic, it is critical that many rinses be used to remove all traces of the mineral from the plant material.

## 6. Hydrogen Peroxide

The concentration of hydrogen peroxide used for surface sterilization of plant material is 30%, ten times stronger than that obtained in a pharmacy. Some researchers have found that hydrogen peroxide is useful for surface-sterilizing material while in the field.

## 7. Enhancing Effectiveness of Sterilization Procedure

» Surfactant (e.g. Tween 20) is frequently added to the sodium hypochlorite.

» A mild vacuum may be used during the procedure.

» The solutions that the explants are in are often shaken or continuously stirred.

## 8. Rinsing

After plant material is sterilized with one of the above compounds, it must be rinsed thoroughly with sterile water. Typically, three to four separate rinses are done.

## 9. Use of Antibiotics and Fungicides in Vitro

We have found that the use of antibiotics and fungicides in vitro is not very effective in eliminating microorganisms and these compounds are often quite phytotoxic.

## 10. Plant Preservative Mixture

PPM™ is a proprietary broad-spectrum biocide, which can be used to control contamination in plant cell cultures, either during the sterilization procedure, or as a medium component. PPM™ comes in an acidic liquid solution (pH 3.8). The recommended dose is 0.5–2.0 mL of PPM™ per liter of medium. Higher doses are required to treat endogenous contamination and for Agrobacterium.

Its makers say that PPM™ has several advantages over antibiotics: It is effective against fungi as well as bacteria, thus it can be substituted for a cocktail of antibiotics and fungicides. PPM™ is less expensive than antibiotics, which makes it affordable for wide and routine use. The formation of resistant mutants toward PPM™ is very unlikely because it targets and inhibits multiple enzymes. Many antibiotics adversely affect plant materials. If used as recommended, PPM™ does not adversely affect in vitro seed germination, callous proliferation, or callous regeneration. Seeds and explants with endogenous contamination can be sterilized at doses of 5-20 mL/L of PPM™. This is useful when routine surface sterilization is insufficient.

## Laboratory Safety and Daily Maintenance Operations

Observance of commonsense safety practices can significantly reduce the possibility of accidents or injuries occurring in a laboratory. For your safety and that of others, observe the following:

> »   Always wear shoes and a laboratory jacket in the laboratory.

> »   Be extremely careful handling alcohols around open flames. They are flammable!

> »   Never pipette by mouth.

> »   Handle hydrochloric acid, sulfuric acid, sodium hydroxide, and other strong acids and alkalis with extreme caution. They are very corrosive!

> »   Wash and bandage all cuts immediately.

> »   Before opening an autoclave, be sure the pressure is reduced to zero and the temperature is below 100°C.

In addition to safety concerns, cleanliness and proper care of equipment are vital to the

operation of a tissue culture laboratory. The following tasks should be performed routinely before the laboratory is closed at night:

1.  Mop floor in lab and culture room with an approved disinfectant.

2.  Turn off hood, unless used continuously to reduce particulates in the air.

3. Fill distilled water tanks and turn off stills.

4. Clean off benches completely and put away chemicals, instruments, glassware, etc.

5. Put microscopes on lowest magnification; turn off and cover them.

6. Shut off all water outlets.

7. Wash and dry all dirty glassware (tubes, pipettes, flasks, etc.).

8. Put away all clean, dry glassware, racks, etc.

9. Turn off all electrical equipment that is not in use overnight (e.g., stirrers, pH meters, balances.)

10. ut away all chemical or media stock solutions.

# Chapter - 7

# Plant Tissue Culture Nutrient Media and Preparation

## Introduction

The methodology of plant tissue culture has advanced to the stage, where tissues from virtually any plant species can be cultured successfully. The successful plant tissue culture depends upon the choice of nutrient medium. The cells of most plant species can be grown on completely defined media. All the media consist of mineral salts, a carbon source (generally sucrose), vitamins and growth regulators. The MS medium designed for tobacco is now used widely for various species, in callus and cell cultures.

## *IN VITRO* MICRO ENVIRONMENT

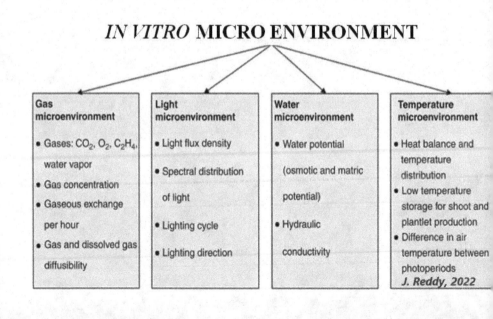

| Gas microenvironment | Light microenvironment | Water microenvironment | Temperature microenvironment |
|---|---|---|---|
| • Gases: $CO_2$, $O_2$, $C_2H_4$, water vapor | • Light flux density | • Water potential (osmotic and matric potential) | • Heat balance and temperature distribution |
| • Gas concentration | • Spectral distribution of light | | • Low temperature storage for shoot and plantlet production |
| • Gaseous exchange per hour | • Lighting cycle | • Hydraulic conductivity | • Difference in air temperature between photoperiods |
| • Gas and dissolved gas diffusibility | • Lighting direction | | |

*J. Reddy, 2022*

*In vitro* Micro-Environment
1. Gases-Oxygen and
   Carbon dioxide
1. Water and Humidity
2. Temperature-Optimal
3. Nutrients-Micro and Macro
4. Light-Artificial

*J. Reddy, 2022*

**In vitro Micro-environment**

Microenvironment is defined as the immediate small-scale environment of a plant cell or tissue, especially as a distinct part of a larger environment, whereas "microenvironmentation" is a process of *in vitro* propagation of plantlets in small culture vessels. Development and growth of plant tissue culture are largely dependent on microenvironmental conditions. The environment of these culture vessels is highly automated and regulated to provide the optimum conditions for the growth of desirable propagules. This process is dependent on the mass and energy exchange process. Water and nutrition are supplied by the medium whereas other optimum conditions such as humidity, temperature, $CO_2$ concentration in light and dark, and $C_2H_4$ concentrations are achieved by several automated processes. Growth and developmental potential of *in vitro* tissue are largely dependent on genes and their level of expression; however, their actual rates are limited by their surrounding environment.

Optimal growth and morphogenesis of tissues may vary for different plants according to their nutritional requirements. Moreover, tissues from different parts of plants may also have different requirements for satisfactory growth. Tissue culture media were first developed from nutrient solutions used for culturing whole plants e.g. root culture medium of White and callus culture medium of Gautheret. White's medium was based on Uspenski and Uspenska's medium for algae. The media used by earlier workers were based on Knop's solution. Subsequently media developed by White (1943) and Heller (1953) were used. Murashige and Skoog's medium (Murashige and Skoog, 1962) is a land mark in plant tissue culture research and is the most frequently used medium for all types of tissue culture work. Based on its composition, other media were evolved to meet the diverse experimental and species-specific requirement. There are – Linsmaier and Skoog (1965), B5 medium of Gamborg et. al. (1968), SH medium of Schenk and Hildebrandt (1972), Nitsch and Nitsch (1969) medium, and woody plant medium (WPM) of Llyod and McCown, (1980).

## Main components of Plant tissue Culture medium

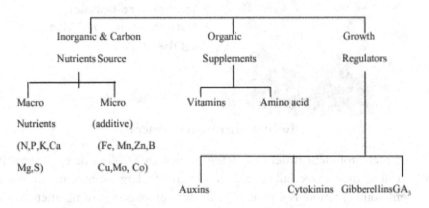

## Media composition

Plant tissue culture media should generally contain some or all of the following components: macronutrients, micronutrients, vitamins, amino acids or nitrogen supplements, source(s) of carbon, undefined organic supplements, growth regulators and solidifying agents. According to the International Association for Plant Physiology the elements in concentrations greater than 0.5 mM.l$^{-1}$ are defined as macroelements and those required in concentrations less than 0.5 mM.l$^{-1}$ as microelements. It should be considered that the optimum concentration of each nutrient for achieving maximum growth rates varies among species.

## Macronutrients

The essential elements in plant cell or tissue culture media include, besides C, H and O, macroelements: nitrogen (N), phosphorus (P), potassium (K), calcium (Ca) magnesium (Mg) and sulphur (S) for satisfactory growth and morphogenesis. Culture media should contain at least 25-60 mM of inorganic nitrogen for satisfactory plant cell growth. Potassium is required for cell growth of most plant species. Most media contain K in the form of nitrate chloride salts at concentrations ranging between 20 and 30 mM. The optimum concentrations of P, Mg, S and Ca range from 1-3 mM if other requirements for cell growth are provided.

## Main components of Plant tissue Culture medium

| Inorganic & Carbon Nutrients Source | Organic Supplements | Growth Regulators |
|---|---|---|

| Macro Nutrients (N,P,K,Ca Mg,S) | Micro (additive) (Fe, Mn,Zn,B Cu,Mo, Co) | Vitamins | Amino acid |

Auxins          Cytokinins  GibberellinsGA$_3$

## Micronutrients

The essential micronutrients (minor elements) for plant cell and tissue growth include iron (Fe), manganese (Mn), zinc (Zn), boron (B), copper (Cu) and molybdenum (Mo). Iron is usually the most critical of all the micronutrients. The element is used as either citrate or tartarate salts in culture media, however, there exist some problems with these compounds for their difficulty to dissolve and precipitate after media preparation. There has been trials to solve this problem by using ethylene diaminetetraacetic acid (EDTA)-iron chelate (FeEDTA). A procedure for preparing an iron chelate solution that does not precipitate have been also developed. Cobalt (Co) and iodine (I) may be added to certain media, but their requirements for cell growth has not been precisely established. Sodium (Na) and chlorine (Cl) are also used in some media, in spite of reports that they are not essential for growth. Copper and cobalt are added to culture media at concentrations of 0.1µM, iron and molybdenum at 1µM, iodine at 5µM, zinc at 5-30 µM, manganese at 20-90 µM and boron at 25-100 µM.

## Carbon and energy sources

In plant cell culture media, besides the sucrose, frequently used as carbon source at a concentration of 2-5%, other carbohydrates are also used. These include lactose, galactose, maltose and starch and they were reported to be less effective than either sucrose or glucose, the latter was similarly more effective than fructose considering that glucose is utilized by the cells in the beginning, followed by fructose. It was frequently demonstrated that autoclaved sucrose was better for growth than filter sterilized sucrose. Autoclaving seems to hydrolyze sucrose into more efficiently utilizable sugars such as fructose. Sucrose was reported to act as morphogenetic trigger in the formation of auxiliary buds and branching of adventitious roots.

It was found that supplements of sugar cane molasses, banana extract and coconut water to basal media can be a good alternative for reducing medium costs. These substrates in addition to sugars, they are sources of vitamins and inorganic ions required growth.

## Vitamins and myo-inositol

Some plants are able to synthesize the essential requirements of vitamins for their growth. Some vitamins are required for normal growth and development of plants, they are required by plants as catalysts in various metabolic processes. They may act as limiting factors for cell growth and differentiation when plant cells and tissues are grown *In vitro*. The vitamins most used in the cell and tissue culture media include: thiamin (B1), nicotinic acid and pyridoxine (B6). Thiamin is necessarily required by all cells for growth. Thiamin is used at concentrations ranging from 0.1 to 10 mg.l$^{-1}$. Nicotinic acid and pyridoxine, however not essential for cell growth of many species, they are often added to culture media. Nicotinic acid is used at a concentration range 0.1-5 mg.l$^{-1}$ and pyridoxine is used at 0.1-10 mg.l$^{-1}$. Other vitamins such as biotin, folic acid, ascorbic acid, pantothenic acid, tocopherol (vitamin E), riboflavin p-amino-benzoic acid are used in some cell culture media however, they are not growth limiting factors. It was recommended that vitamins should be added to culture media only when the concentration of thiamin is below the desired level or when the cells are required to be grown at low population densities. Although it is not a vitamin but a carbohydrate, myo-inositol is added in small quantities to stimulate cell growth of most plant species. Myo-inositol is believed to play a role in cell division because of its breakdown to ascorbic acid and pectin and incorporation into phosphoinositides and phosphatidyl-inositol. It is generally used in plant cell and tissue culture media at concentrations of 50-5000 mg.l$^{-1}$.

## Amino acids

The required amino acids for optimal growth are usually synthesized by most plants however, the addition of certain amino acids or amino acid mixtures is particularly important for establishing cultures of cells and protoplasts. Amino acids provide plant cells with a source of nitrogen that is easily assimilated by tissues and cells faster than inorganic nitrogen sources. Amino acid mixtures such as casein hydrolysate L-glutamine, L-asparagine and adenine are frequently used as sources of organic nitrogen in culture media. Casein hydrolysate is generally used at concentrations between 0.25-1 g.l⁻¹. Amino acids used for enhancement of cell growth in culture media included; glycine at 2 mg.l⁻¹, glutamine up to 8 mM, asparagine at 100mg.l⁻¹, L-arginine and cysteine at 10 mg.l⁻¹ and L-tyrosine at 100mg.l⁻¹.

## Undefined organic supplements

Some media were supplemented with natural substances or extracts such as protein hydrolysates, coconut milk, yeast extract, malt extract, ground banana, orange juice and tomato juice, to test their effect on growth enhancement. A wide variety of organic extracts are now commonly added to culture media. The addition of activated charcoal is sometimes added to culture media where it may have either a beneficial or deleterious effect. Growth and differentiations were stimulated in orchids, onions and carrots, tomatoes. On the other hand, an inhibition of cell growth was noticed on addition of activated charcoal to culture medium of soybean. Explanation of the mode of action of activated charcoal was based on adsorption of inhibitory compounds from the medium, adsorption of growth regulators from the culture medium or darkening of the medium. The presence of 1% activated charcoal in the medium was demonstrated to largely increase hydrolysis of sucrose during autoclaving which cause acidification of the culture medium.

## Solidifying agents

Hardness of the culture medium greatly influences the growth of cultured tissues. There are a number of gelling agents such as agar, agarose and gellan gum. Agar, a polysaccharide obtained from seaweeds, is of universal use as a gelling agent for preparing semi-solid and solid plant tissue culture media. Agar has several advantages over other gelling agents; mixed with water, it easily melts in a temperature range 60-100°C and solidifies at approximately 45°C and it forms a gel stable at all feasible incubation temperatures. Agar gels do not react with media constituents and are not digested by plant enzymes. It is commonly used in media at concentrations ranging between 0.8-1.0%. Pure agar preparations are of great importance especially in experiments dealing with tissue metabolism. Agar contains Ca, Mg and trace elements on comparing different agar brands; Bacto, Noble and purified agar, in concern with contaminants. The author, for example reported Bacto agar to contain 0.13, 0.01, 0.19, 0.43, 2.54, 0.17% of Ca, Ba, Si, Cl, SO4⁻, N, respectively. Impurities also included 11.0, 285.0 and 5.0 mg.l⁻¹ for iron, magnesium and copper as contaminants, respectively. Amounts of some contaminants were higher in purified agar than in Bacto agar of which Mg that accounted for 695.0 mg.l⁻¹ and Cu for 20.0 mg.l⁻¹.

Reduction of culture media costs is continually targeted in large-scale cultures and search for cheap alternatives provided that white flower, potato starch, rice powder were as good gelling agents as agar. It was also experienced that combination of laundry starch, potato starch and semolina in a ratio of 2:1:1 reduced costs of gelling agents by more than 70%.

## Growth regulators

Plant growth regulators are important in plant tissue culture since they play vital roles in stem elongation, tropism, and apical dominance. They are generally classified into the following groups; auxins, cytokinins, gibberellins and abscisic acid. Moreover, proportion of auxins to cytokinins determines the type and extent of organogenesis in plant cell cultures.

## Auxins

The common auxins used in plant tissue culture media include: indole-3- acetic acid (IAA), indole-3- butric acide (IBA), 2,4-dichlorophenoxy-acetic acid (2,4-D) and naphthalene- acetic acid (NAA). IAA is the only natural auxin occurring in plant tissues There are other synthetic auxins used in culture media such as 4-chlorophenoxy acetic acid or p-chloro-phenoxy acetic acid (4-CPA, pCPA), 2,4,5-trichloro-phenoxy acetic acid (2,4,5 T), 3,6-dichloro-2-methoxy- benzoic acid (dicamba) and 4- amino-3,5,6-trichloro-picolinic acid (picloram).

Auxins differ in their physiological activity and in the extent to which they translocate through tissue and are metabolized. Based on stem curvature assays, eight to twelve times higher activity was reported on using 2,4-D than IAA, four times higher activity of 2,4,5 T in comparison with IAA and NAA has as doubled activity as IAA. In tissue cultures, auxins are usually used to stimulate callus production and cell growth, to initiate shoots and rooting, to induce somatic embryogenesis, to stimulate growth from shoot apices and shoot stem culture. The auxin NAA and 2,4-D are considered to be stable and can be stored at 4°C for several months. The solutions of NAA and 2,4-D can also be stored for several months in a refrigerator or at -20°C if storage has to last for longer periods. It is best to prepare fresh IAA solutions each time during medium preparation, however IAA solutions can be stored in an amber bottle at 4°C for no longer than a week. Generally IAA and 2,4-D are dissolved in a small volume of 95% ethyl alcohol. NAA, 2,4-D and IAA can be dissolved in a small amount of 1N NaOH. Chemical structures of some of the frequently used auxins are given in figures below. There are also some auxinlike compounds that vary in their activity and are rarely used in culture media.

## Cytokinins

Cytokinins commonly used in culture media include BAP (6-benzyloaminopurine), 2iP (6-dimethylaminopurine), kinetin (N-2-furanylmethyl-1H-purine-6-amine), Zeatin (6-4-hydroxy-3-methyl-trans-2-butenylaminopurine) and TDZ (thiazuron-N-phenyl-N-1,2,3 thiadiazol-5ylurea). Zeatin and 2iP are naturally occurring cytokinins and zeatin is more effective. In culture media, cytokinins proved to stimulate cell division, induce shoot formation and axillary shoot proliferation and to retard root formation. The cytokinins are relatively stable compounds in culture media and can be stored desiccated at -20°C. Cytokinins are frequently reported to be difficult to dissolve and sometimes addition of few drops of 1N HCl or 1N NaOH facilitate their dissolution. Cytokinins can be dissolved in small amounts of dimethylsulfoxide (DMSO) without injury to the plant tissue. DMSO has an additional advantage because it acts as a sterilizing agent; thus stock solutions containing DMSO can be added directly to the sterile culture medium. Chemical structure of the frequently used in plant tissue culture media is given below.

## Gibberellins

Gibberellins comprise more than twenty compounds, of which GA3 is the most frequently used gibberellin. These compounds enhance growth of callus and help elongation of dwarf plantlets.

Other growth regulators are sometimes added to plant tissue culture media as abscisic acid, a compound that is usually supplemented to inhibit or stimulate callus growth, depending upon the species. It enhances shoot proliferation and inhibits later stages of embryo development. Although growth regulators are the most expensive medium ingredients, they have little effect on the medium cost because they are required in very small concentrations.

A comparison of the chemical composition of the frequently used plant tissue culture media appears in the Table (1) which was given in the appendix of the proceedings of the technical meeting of the International Atomic Energy Agency (IAEA).

# Table 1. Composition of media most frequently used.

| Medium / Components (mg.l$^{1-}$) | MS | G5 | W | LM | VW | Km | M | NN |
|---|---|---|---|---|---|---|---|---|
| Macronutrients | | | | | | | | |
| Ca3(PO4)2 | | | | | 200.0 | | | |
| NH4NO3 | 1650.0 | | | 400.0 | | | | 720.0 |
| KNO3 | 1900.0 | 2500.0 | 80.0 | | 525.0 | 180.0 | 180.0 | 950.0 |
| CaCl2.2H2O | 440.0 | 150.0 | | 96.0 | | | | 166.0 |
| MgSO4.7H2O | 370.0 | 250.0 | 720.0 | 370.0 | 250.0 | 250.0 | 250.0 | 185.0 |
| KH2PO4 | 170.0 | | | 170.0 | 250.0 | 150.0 | 150.0 | 68.0 |
| (NH4)2SO4 | | 134.0 | | | 500.0 | 100.0 | 100.0 | |
| NaH2PO4.H2O | | 150.0 | 16.5 | | | | | |
| CaNO3.4H2O | | | 300.0 | 556.0 | | 200.0 | 200.0 | |
| Na2SO4 | | | 200.0 | | | | | |
| KCl | | | 65.0 | | | | | |
| K2SO4 | | | | 990.0 | | | | |
| Micronutrients | | | | | | | | |
| KI | 0.83 | 0.75 | 0.75 | | | 80.0 | 0.03 | |
| H3BO3 | 6.20 | 3.0 | 1.5 | 6.2 | | 6.2 | 0.6 | 10.0 |
| MnSO4.4H2O | 22.30 | | 7.0 | | 0.75 | 0.075 | | 25.0 |
| MnSO4.H2O | | 10.0 | | 29.43 | | | | |
| ZnSO4.7H2O | 8.6 | 2.0 | 2.6 | 8.6 | | | 0.05 | 10.0 |
| Na2MoO4.2H2O | 0.25 | 0.25 | | 0.25 | | 0.25 | 0.05 | 0.25 |
| CuSO4.5H2O | 0.025 | 0.025 | | 0.25 | | 0.025 | | 0.025 |
| CoCl2.6H2O | 0.025 | 0.025 | | | | 0.025 | | |
| Co(NO3)2.6H2O | | | | | | | 0.05 | |
| Na2EDTA | 37.3 | 37.3 | | 37.3 | | 74.6 | 37.3 | 37.3 |
| FeSO4.7H2O | 27.8 | 27.8 | | 27.8 | | 25.0 | 27.8 | 27.8 |
| MnCl2 | | | | | | 3.9 | 0.4 | |
| Fe(C4H4O6)3.2H2O | | | | | 28.0 | | | |
| Vitamins and other supplements | | | | | | | | |
| Inositol | 100.0 | 100.0 | | 100.0 | | | | 100.0 |
| Glycine | 2.0 | 2.0 | 3.0 | 2.0 | | | | 2.0 |
| Thiamine HCl | 0.1 | 10.0 | 0.1 | 1.0 | | 0.3 | 0.3 | 0.5 |
| Pyridoxine HCl | 0.5 | | 0.1 | 0.5 | | 0.3 | 0.3 | 0.5 |

| Nicotinic acid | 0.5 | | 0.5 | 0.5 | | | 1.25 | 5.0 |
|---|---|---|---|---|---|---|---|---|
| Ca-panthothenate | | | 1.0 | | | | | |
| Cysteine HCl | | | 1.0 | | | | | |
| Riboflavin | | | | | | 0.3 | 0.05 | |
| Biotin | | | | | | | 0.05 | 0.05 |
| Folic acid | | | | | | | 0.3 | 0.5 |

MS Murashige and Skoog, G5= Gamborg *et al.*, W= White, LM= Lloyd and McCown, VW= Vacin and Went, Km= Kudson modified, M= Mitra *et al.* and NN= Nitsch and Nitsch media.

## Media preparation

Preparation of culture media is preferred to be performed in an equipped for this purpose compartment. This compartment should be constructed so as to maintain ease in cleaning and reducing possibility of contamination. Supplies of both tap and distilled water and gas should be provided. Appropriate systems for water sterilization or deionization are also important. Certain devices are required for better performance such as a refrigerator, freezer, hot plate, stirrer, pH meter, electric balances with different weighing ranges, heater, Bunsen burner in addition to glassware and chemicals. It is well known now that mistakes which occur in tissue culture process most frequently originate from inaccurate media preparation that is why clean glassware, high quality water, pure chemicals and careful measurement of media components should be facilitated.

A convenient method for preparation of culture media is to make concentrated stock solutions which can be immediately diluted to preferred concentration before use. Solutions of macronutrients are better to be prepared as stock solutions of 10 times the strength of the final operative medium. Stock solutions can be stored in a refrigerator at 2- 4°C. Micronutrients stock solutions are made up at 100 times of the final concentration of the working medium. The micronutrients stock solution can also be stored in a refrigerator or a freezer until needed. Iron stock solution should be 100 times concentrated than the final working medium and stored in a refrigerator. Vitamins are prepared as either 100 or 1000 times concentrated stock solutions and stored in a freezer (-20°C) until used if it is desired to keep them for long otherwise they can be stored in a refrigerator for 2-3 months and should be discarded thereafter. Stock solutions of growth regulators are usually prepared at 100-1000 times the final desired concentration.

Concentrations of inorganic and organic components of media are generally expressed in mass values (mg.l$^{-1}$, mg/l and p.p.m.) in tissue culture literature. The International Association for plant Physiology has recommended the use of mole values. Mole is an abbreviation for gram molecular weight which is the formula weight of a substance in grams. The formula weight of a substance is equal to the sum of weights of the atoms in the chemical formula of the substance. One liter of solution containing 1mole of a substance is 1 molar (1M) or 1 mol.l$^{-1}$ solution of the substance (1 mol.l$^{-1}$= $10^3$ mmol.l$^{-1}$= $10^6$ µmol/l). It is routinely now to accepted to express concentrations of macronutrients and organic nutrients in the culture medium as mmol/l values, and µmol/l values for micronutrients, hormones, vitamins and other organic constituents. This was explained on the basis that mole values for all compounds have constant number of molecules per mole.

## Media selection

For the establishment of a new protocol for a specific purpose in tissue culture, a suitable medium is better formulated by testing the individual addition of a series of concentrations of a given compound to a universal basal medium such as MS, LS or B5. The most effective variables in plant tissue culture media are growth regulators, especially auxins and cytokinins. Full strength of salts in media proved good for several species, but in some species the reduction of salts level to ½ or ¼ the full concentration gave better results in *in vitro* growth.

Sucrose is often assumed to be the best source of carbon for in vitro culture, the levels used are from 2 to 6% and the level has to be defined for each species.

## Media sterilization

Prevention of contamination of tissue culture media is important for the whole process of plant propagation and helps to decrease the spread of plant parasites. Contamination of media could be controlled by adding antimicrobial agents, acidification or by filtration through microporous filters. To reduce possibilities of contamination, it is recommended that sterilization rooms should have the least number of openings. Media preparation and sterilization are preferred to be performed in separate compartments. Sterilization area should also have walls and floor that withstand moisture, heat and steam.

Sterilization of media is routinely achieved by autoclaving at the temperature ranging from 115° – 135° C. Advantages of autoclaving are: the method is quick and simple, whereas disadvantages are the media pH changes and some components may decompose and so to loose their effectiveness. As example autoclaving mixtures of fructose, glucose and sucrose resulted in a drop in the agar gelling capacity and affecting pH of the culture medium through the formation of furfural derivatives due to sucrose hydrolysis.

Filtration through microporus filters (0.22- 0.45) is also used for thermolabile organic constituents such as vitamins, growth regulators and amino acids. Additives of antimicrobial agents are less commonly applied in plant tissue culture media. Limitation for their use was reported and attributed to harm imposed on plants as well.

## Plant Growth Regulators Commonly Used in Plant Tissue Culture

| Class | Name | Abbreviation | Comments |
|---|---|---|---|
| Auxin | Indole-3-acetic acid | IAA | Use for callus induction at 10-30 μM. Lowering to 1-10 μM can stimulate organogenesis. Is inactivated by light and readily oxidized by plant cells. The synthetic auxins below have largely superceded IAA for tissue culture studies. |
| | Indole-3-butyric acid | IBA | Use for rooting shoots regenerated via organogenesis. Either maintain at a low concentration (1-50 μM) throughout the rooting process, or expose to a high concentration (100-250 μM) for 2-10 days and then transfer to hormone-free medium. Can also use as a dip for *in vitro* or *ex vitro* rooting of shoots. |
| | 2,4-Dichlorophenoxyacetic acid | 2,4-D | Most commonly used synthetic auxin for inducing callus and maintaining callus and suspension cells in dedifferentiated states. Usually used as sole auxin source (1-50 μM), or in combination with NAA. |
| | p-Chlorophenoxyacetic acid 1-Naphthaleneacetic acid | pCPA NAA | Similar to 2,4-D, but less commonly used. Synthetic analogue of IAA. Commonly used either as sole auxin source (2-20 μM for callus induction and growth of callus and suspension cultures; 0.2-2 μM for root induction), or in combination with 2,4-D. |

| Cytokinin | 6-Furfurylaminopurine (kinetin) | K | Often included in culture media for callus induction, growth of callus and cell suspensions, and induction of morphogenesis (1-20 μM). Higher concentrations (20-50 μM) can be used to induce the rapid multiplication of shoots, axillary/adventitious buds, or meristems. |
|---|---|---|---|
| | 6-Benzylaminopurine | BAP, BA | Included in culture media for callus induction, growth of callus and cell suspensions (0.5-5.0 μM), and for induction of morphogenesis (1-10 μM). More commonly used than kinetin for inducing rapid multiplication of shoots, buds, or meristems (5-50 μM). |
| | $N$-Isopentenylaminopurine | 2iP | Less commonly used than K or BAP for callus induction and growth (2-10 μM), induction of morphogenesis (10-15 μM), or multiplication of shoots, buds, or meristems (30-50 μM). |
| | Zeatin | Zea | Seldom used in callus or suspension media. Can be used for induction of morphogenesis (0.05-10 μM). Zea is thermolabile and must not be autoclaved. |
| | Thidiazuron | TDZ | Thidiazuron ($N$-phenyl-$N'$-1,2,3-thiadiazol-5-ylurea) is first reported to have cytokinin activity since 1982. It has been used successfully in vitro to induce adventitious shoot formation and to promote axillary shoot proliferation. It is especially effective with recalcitrant woody species. Low concentrations of thidiazuron (0.0022 to 0.088 mg/liter) are also very effective. |
| Gibberellin | Gibberellin $A_3$ | $GA_3$ | Seldom used in callus or suspension medium (one exception being potato). Can promote shoot growth when added to shoot induction medium at 0.03-14 μM. Also used to enhance development in embryo/ovule cultures (0.3-48 μM). $GA_3$ is thermolabile and must not be autoclaved. |
| Abscisic acid | Abscisic acid | ABA | Used at concentrations of 0.4-10 μM to prevent precocious germination, and promote normal development of somatic embryos. |

# Chapter - 8

# Types of Plant Tissue Culture-Organ Culture

Basically, any living part of the plant can used as an expant.

**Following are the common types of plant tissue culture:**

1. Cell or suspension culture
2. Protoplast culture
3. Organ culture
4. Callus culture
5. Embryo culture
6. Anther and pollen culture
7. Ovule culture
8. Ovary culture

But the most important and widely used type is organ culture. There are several types of organ cultures and the most common types are meristematic culture, shoot tip, nodal culture of separate lateral bud, isolated root, and embryo culture.

## Organ Cultures

In PTC organ culture refers to the in vitro culture and maintenance of an excised organ primordia or whole or part of an organ in a way that may allow differentiation and preservation of the architecture and/or function. W. Kotte and W. J. Robbins in 1922 reported first the culture of excised root tips from the aseptically germinated wheat seedlings. Practically each and every living part or organ of the plant can be cultured to produce plants. This is possible due to totipotency of plants and their cells.

## Flower Culture

Flower culture can be defined as the aseptic culture of excised floral bud on a chemically defined nutrient medium where they continue their development to produce a full bloom in a culture vessel. Young and complete flower culture can also be described as flower culture. In culture medium, the flowers remain healthy and they develop normally to mature seeds.

**Principle:** Flowers can be cultured at the different stages of development, such as primordial stage, bud stage, pre-pollination stage and post-pollination stage. Flower primordia and the young flower bud require a complex medium containing inorganic salts, B-vitamins, amino acids, coconut milk, auxins and cytokinins. The mature flowers at pre or post-pollination stage need comparatively simple media containing inorganic salts, sucrose and a small quantity of hormones.

**Protocol:**

1. Flower buds or mature flowers are collected from the healthy plants.

2. Wash them thoroughly and dip them in 5% Teepol solution for 10 minutes and wash.

3. Transfer them to laminar air-flow cabinet. Surface sterilizes them by immersing in 5% Sodium hypochlorite, wash with autoclaved distilled water.

4. Using flamed forceps, transfer the flower bud or mature flower to culture tubes containing 20 ml solid medium.

5. Incubate the culture in 16 hrs of light at 25° C.

**Importance:**

1. The main application of floral primordia or flower bud culture is in fundamental studies of flower development.

2. Flowers put into the culture before pollination do not usually produce fruits. In some cases, parthenocarpic fruit development has been observed, particularly in presence of auxins.

3. The culture of pollinated flowers is very important to study the fruit

development. Often the in vitro fruits are smaller than their natural counterparts, but the size can be increased by supplementing the medium with an appropriate combination of growth hormones such as auxins, gibberellins and cytokinins.

4. Flower culture has been used to study the sex expression in flower. In the cucumber (Cucumis sativus), there exist different genetic lines that are monoecious (with unisexual male or female flowers on the same plant), gynoecious (purely female) or hermaphrodite.

Under suitable natural conditions, the monocious types will produce only staminate male flower and the gynoecious types only pistilated female flower.

It has been observed that in culture the potentially male buds tend to form ovaries and this tendency is promoted when IAA is added to the culture medium. In contrast, addition of gibberellic acid counteracts the effect of auxin. Isolated potentially female or bisexual flower buds in culture remain unchanged even in presence of IAA or gibberellic acid or cytokinins. Such culture techniques are also important for experimental studies on floral morphogenesis.

## Shoot Tip/Meristem Culture

Shoot tip culture may be described as the culture of terminal (0.1-1.0 mm) portion of a shoot comprising the meristem (0.05-0.1 mm) together with primordial and developing leaves and adjacent stem tissue.

**Meristem Culture:** Meristem culture is the in vitro culture of a generally shiny special dome-like structure measuring less than 0.1 mm in length and only one or two pairs of the youngest leaf primordia, most often excised from the shoot apex.

**Mericloning:** Mericloning is a popular term. It is not applied in scientific usage. It refers to the in vitro vegetative propagation of orchids from excised shoot tips, axillary buds or floral organs.

**Meristemming:** Meristem Ming is also a popular term. It is used to describe the in vitro clonal propagation of plants from various explant sources including shoot tips, leaf sections and calli.

**Principle:** The excised shoot tips and meristem can be cultured aseptically on agar solidified simple nutrient medium or on paper bridges dipping into liquid medium and under the appropriate condition will grow out directly into a small leafy shoot or multiple shoots. Alternatively the meristem may form a small callus at its cut case on which a large number of shoot primordia will develop.

These shoot primordia grow out into multiple shoots. Once the shoots have been grown directly from the excised shoot tip or meristem, they can be propagated further by nodal cuttings. This process involves separating the shoot into small segments each containing one node. The axillary bud on each segment will grow out in culture to form yet another shoot.

The excised stem tips of orchids in culture proliferate to form callus from which some organised juvenile structures known as proto-corm develop. When the proto-corms are separated and cultured to fresh medium, they develop into normal plants. The stem tips of Cascuta reflexa in culture can be induced to flower when they are maintained in the dark. Exogenously supplied cytokinins in the nutrient medium plays a major role for the development of a leafy shoot or multiple shoots from meristem or shoot tip. Generally, high cytokinin and low auxin are used in combination for the culture of shoot tip or meristem.

Addition of adenine sulfate in the nutrient medium also induces the shoot tip multiplication in some cases. BAP is the most effective cytokinin commonly used in shoot tip or meristem culture. Similarly, NAA is the most effective auxin used in shoot tip culture. Coconut milk and gibberellic acid are also equally effective for the growth of shoot apices in some cases.

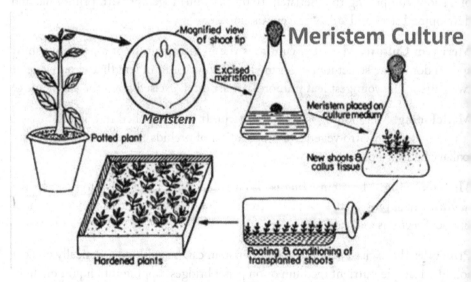

**Protocol:**

1.  Remove the young twigs from a healthy plant. Cut the tip (1 cm) portion of the twig.

2.  Surface sterilize the shoot apices by incubation in a sodium hypochlorite solution (1% available chlorine) for 10 minutes. The ex- plants are thoroughly rinsed 4 times in sterile distilled water.

3.  Transfer each explant to a sterilized petri dish.

4.  Remove the outer leaves from each shoot apices with a pair of jeweler's forceps. This lessens the possibility of cutting into the softer underlying tissues.

5.  After the removal of all outer leaves, the apex is exposed. Cut off the ultimate apex with the help of scalpel and transfer only those less than 1 mm in length to the surface of the agar medium or to the surface of filter-paper Bridge. Flame the neck of the culture tube before and after the transfer of the excised tips. Binocular dissecting microscope can be used for cutting the true meristem or shoot tip perfectly.

6.  Incubate the culture under 16hrs light at 25°C.

7.  As soon as the growing single leafy shoot or multiple shoots obtained from single shoot tip or meristem, develop root, transfer them to hormone free medium.

8.  The plantlets formed by this way are later transferred to pots containing compost and kept under greenhouse conditions.

**Importance of Shoot Tip/Meristem Culture**

**The uses of shoot tips and meristem in tissue culture are very varied and include mainly:**

(i) Virus eradication,

(ii) Micro-propagation and

(iii) Storage of genetic resources.

**Virus Eradication:**

Many important plants contain systemic viruses which substantially reduce their potential yield and quality. It is, therefore, important to produce virus free stocks which can be multiplied. Generally, highly meristematic tissue of a virus infected plant remains free from virus due to their fast mitotic activity. Therefore, shoot tips and meristems of a virus infected plant are the ideal explants to produce a virus free stock. This technique is also valuable for the maintenance of carefully defined stocks of specific varieties and cultivars in disease Free State. The size of the meristem explant is critical for virus eradication. Often so called meristem tip cultures have failed to eliminate virus infection because the explant contains shoot apices with vascular tissue instead of true meristem.

This technique, combined with heat treatment (thermotherapy) or chemical treatment (chemotherapy) has proved to be very effective in virus eradication. Heat treatment is done by placing an actively growing plant in a thermotherapy chamber. Over a period of two weeks the temperature is increased to 38°C inside the chamber and the plants are maintained at this temperature for two months.

After that period, the apical meristem is excised, surface sterilized and transferred aseptically to agar medium. Using this technique 85% to 90% virus free plants have been obtained. Without heat treatment, shoot tips or meristem can be grown on chemotherapeutants added medium for virus eradication. Commonly used chemotherapeutants are 2, 4-D, mela-chife green, thiouracil etc.

Shoot tip or meristem culture has enormous horticultural value e.g. in the production of plants for the cut flower industry when stock plants of registered lines must be maintained in as near-perfect condition as possible. Any infection by virus that affect the growth or physical characteristics of size and shape of flowers is obviously very serious problem from commercial point of view.

Meristem culture technique to clean up the stocks could, therefore, avert a commercial disaster. Similarly, in the agricultural world, the production or yield of a crop can fall dramatically as a result of a viral infection and render that particular variety no longer salable or commercial value. Meristem culture could be of value in restoring the original properties of the variety by removing the infection.

**Micropropagation:**

A sexual or vegetative propagation of whole plants using tissue culture techniques is referred to as micro-propagation. Shoot tip or meristem culture of many plant species can successfully be used for micro-propagation.

## Storage of Genetic Resources:

Many plants produce seeds that are highly heterozygous in nature or that is recalcitrant. Such seeds are not accepted for storing genetic resources. So, the meristem from such plants can be stored in vitro. Besides the above-mentioned uses of shoot tip or meristem culture, it can also be utilized in various important fields of plant science such as:

## Shoot Tip or Meristem Culture and Plant Breeding:

In many plant breeding experiments, the hybrid plants produce abortive seeds or nonviable seeds. As a result, it makes a barrier to crossibility in plants where non-viable seeds are unable to develop into mature hybrid plants. Shoot tip or meristem from such hybrid plant can be cultured to speed up breeding programme.

## Propagation of Haploid Plants:

Haploid plants derived from anther or pollen culture always remain sterile unless and until they are made homozygous diploid. Meristem or shoot tip culture of haploid plants can be used for their propagation from which detailed genetic analysis can be done on the basis of morphological characters and biochemical assay.

## Meristem or Shoot Tip Culture and Quarantine:

There are some strict regulations concerning the international movement of vegetative plant material. After thoroughly checking, the plant materials may be rejected by quarantine authority if the plant material carries some diseases.

Plantlets derived from shoot tip or meristem culture are easily accepted by the quarantine authority for international exchange without any checking. Therefore, using this technique, crop plants can be easily exchanged in crop improvement programmes that are based on materials from different parts of the world.

## Leaf Culture

Leaf culture is the culture of excised young leaf primordia or immature young leaf of the shoot apex in a chemically defined medium where they grow and follow the developmental sequences under controlled conditions.

**Principle:**

Leaf primordia or very young leaves are excised, surface sterilized and inoculated on an agar solidified medium. In culture leaf remains in healthy condition for a long period. Leaves can be taken from aseptically grown plants for culture. Since leaves have a limited growth potential, so in culture the amount of leaf growth depends upon the stage of maturity at the time of excision. Leaf primordia or very young leaf have more growth potential than nearly mature leaves.

Most of the work on leaf culture has been done with lower plants, particularly fern (Osmunda), although higher plant species, such as tobacco and sunflower, have been used. In culture, the fern leaf primordia (1.2 mm), excised from underground buds, develop into leaves having a normal morphology except that they are much reduced in size than in vivo leaves due to a reduced number of cells rather than a decrease in cell size. The growth of cultured leaf primordia is also completed earlier than intact leaf.

It has also been found that there is a correlation between leaf primordia size and its mode of development in culture. In Osmunda cinnamomea, smallest leaf primordia (300 μm in length) give rise to shoots instead of leaf in culture. However, with increasing size of primordia, there is an increased tendency to form leaves. These results indicate that some unidentified leaf forming substances gradually accumulate as the primordia develop.

**Protocol:**

1.  Detach vegetative bud or very young leaf from shoot apex at the vegetative phase of the plant. Wash the explants thoroughly with running tap water.

2.  Immerse the leaf buds or young leaves in 5% Teepol for 10 minutes. Wash the explants to remove Teepol.

3.  Leaf buds or young leaves are surface sterilized by immersion in 70% v/v Ethanol for 30 seconds. This treatment is followed by 10-15 minutes incubation in sodium hypochlorite solution with 0.8% available chlorine. Rinse the explants 3-4 times in sterile distilled water.

4.  Excise the leaf primordia from the leaf bud with the help of surgical scalpel.

5.  Inoculate the leaf primordia or young leaf onto 20 ml of solidified medium in a culture tube.

5. Incubate the culture at 25° C under 16 hrs. light.

## Importance of Leaf Culture

1. Culture of excised leaf primordia is valuable to study the effects of various nutrients, growth factors and changing environmental conditions on leaf development under conditions divorced from the complexities of the intact plant.

2. In case of fern, leaf primordia cultures are used to study the formation of sporangia and the size at which a primordium is destined to become a leaf.

3. Young leaves of most of the solanaceous species form numerous shoot buds instead of callus formation when they are cultured in solidified MS medium supplemented with 1-5 μm kinetin or BAP or 2iPA. When shoot have grown to a height 3 cm, they may be removed and sub cultured onto MS medium devoid of growth hormones. Root formation is stimulated by this treatment. Therefore, leaf culture of solanaceous species can be used as clonal micro-propagation.

4. The lid is quickly replaced and the whole dish is swirled gently to disperse the cell and medium mixture uniformly throughout the lower half of the petri dish.

5. The medium is allowed to solidify and the petri dish is kept at the inverted position.

6. The cultures are incubated under 16hrs light (3,000 lux, cool white) or under continuous dark at 25°C.

7. The petri dishes are observed at regular intervals under inverted microscope to see whether the cells have divided or not.

8. After certain days of incubation, when the cells start to divide, a grid is drawn on the undersurface of the petri dish to facilitate counting the number of dividing cells.

9. The dividing cells ultimately form pin-head shaped cell colonies within 21 days of incubation.

10. The plating efficiency (PE) can be calculated from the counting of cell colonies by the following formula:

   PE = Number of colonies per plate/Number of total cell per plate x 100

11. Pin-head shaped colonies, when they reach a suitable size, are transferred to fresh medium for further growth.

# Root Culture

Root culture can be defined as the culture of excised radical tips of aseptically germinated seeds in a liquid medium where they are induced to grow independently under controlled conditions.

## Principle:

Intact in vivo plants are not suitable for the isolation of intact root tips because the roots of 'the plant are buried deeply in the soil. Again, root tips from young seedlings are very sensitive to toxic sterilants. So it is better to avoid the surface sterilization of young root tips for the establishment of root cultures. Root cultures can be successfully initiated from the excised radicle tips of aseptically germinated seeds.

Root tip cultures are generally maintained in moving liquid medium. In culture, root tips are induced to grow like that of root system of an intact plant. A clone of excised roots can also be established from a single root culture by repeatedly cutting and transferring of the main root tips or of lateral tips into fresh medium in every subculture at the interval of definite period. Growth of excised roots can be expressed in terms of fresh and dry weight, increase in length of the main axis, number of emergent laterals and total length of laterals per culture.

## Protocol:

### Initiation of Isolated Root Culture:

(1) Seeds are surfaced sterilized by the conventional methods and germinated on moist filter paper or White's basal medium at 25°C in the dark (Fig 2.1).

(2) When the seedling roots are 20 to 40 mm in length, 10 mm apical tips (tip inoculum) are excised with a scalpel and each transferred to 40 ml of liquid medium contained in 100 ml wide-necked Erlenmeyer flasks.

(3) Flasks are incubated at 25°C in the dark.

### Initiation of Clones:

The root material derived from a single radicle tip could be multiplied and maintained in continuous culture. Such genetically uniform root cultures are referred to as a clone of isolated roots. Initiation of root clones is a very simple technique.

**The protocol is given below:**

1. Establish a root culture from a radical tip of a seed as described above.

2. Transfer a 10-day-old established root culture to a sterile petri dish containing sterile medium. Next, using flamed scissors, cut the main axis of root into a number of pieces (each piece is called sector inoculum or initial), each bearing four or five young laterals.

3. Transfer the individual sector inoculum aseptically to a flask liquid medium and incubate in dark at 25°C.

4. Such sector culture can be used to initiate further tip culture using 10mn apical tips of laterals of a growing sector inoculum or the growing sector is again cut into 4-5 sectors to initiate the sector culture.

## Importance of Root Culture

The root of many species cannot be cultured. Studies with successful species (Tomato, Pea, Clover, Carrot, etc.) have contributed a lot of significant information's.

**The importance of root culture is given below:**

**Importance of Root Culture in Relation to Basic Information:**

1. Root cultures have increased our knowledge of carbohydrate metabolism, role of mineral ions, vitamins etc. in root growth.

2. Root cultures have provided basic information regarding the dependence of roots on shoots for growth hormones.

3. Root clones are ideally suited for the study of the effect of various compounds on root growth.

**Specific Applications of Root Culture:**

**Study of Nodulation of Leguminous Root in Culture:**

The process of nodule formation on the roots of leguminous plant by the nitrogen-fixing (NIF) bacteria (Rhizobium sp.) is a complex physiological system which is poorly understood. Root cultures of leguminous plant provide an ideal system to study it. When bacteria are inoculated directly in root culture, they quickly grow and spoil the whole culture.

Again, nitrate in the medium is required for the root growth but is inhibitory to nodulation. To overcome such difficulties M Raggio, N Raggio and J G Torrey modified the root culture technique for nodulation study. In their study, the base of an excised root of Phaseolus vulgaris was supplied with sucrose and vitamins via agar medium in a glass vial.

The remainder of the root was in contact with an inorganic nitrate-free medium containing Rhizobium. By this process, isolated roots develop nodules in culture. Therefore, in vitro nodulation helps to understand the relationship between symbiotic NIF bacteria and higher plants.

**Regeneration of Shoots on Roots:**

Culture of isolated roots can be maintained continuously for many years. However, in some species e.g. Atropa, Convolvulus arvensis, shoots can be induced to regenerate from cultured roots. The shoot primordia can be derived from callus at the cut ends of the roots, as in case of Atropa or endogenously from the internal tissues of the root as in Convolvulus. This phenomenon is of practical value as well as theoretical interest.

**Study of Synthesis of Secondary Metabolites from Root Culture:**

The roots of many medicinally important plant species synthesize pharmaceutical important alkaloids as by-products of normal metabolism (secondary metabolites). Root culture have been used to locate the site of biosynthesis of such compounds. Root culture techniques are also used to increase the synthesis of such compounds in cultured root by some nutritional manipulations.

**Initiation and Development of Secondary Vascular Tissues:**

Normally excised cultured roots show only the primary structure of young seedling radicle and, therefore, do not form secondary vascular tissue. But it has been found that excised root tips from pea seedling develop a vascular cambium when cultured in medium containing indoleacetic acid. Attempts have also been made to define the factors that determine the site, time of origin and functioning of such vascular cambium.

Torrey studied this phenomenon using a modification of the technique developed by Raggio and Raggio. The basal 5 mm portions of excised 15 mm long pea root were inserted into an agar solidified basal medium supplemented with various test substances e.g., auxins, cytokinins, mesoinositol and extra sucrose and the exposed 10 mm portion of the root was placed in a petri dish of basal agar medium containing inorganic salts, vitamins and sucrose.

When the bases of pea roots were fed with extra sucrose (8%) and IAA ($10^{-7}$M) a vascular cambium was initiated. But when the whole portion of roots was allowed to grow in a basal medium, they did not form secondary vascular tissues. These experiments have suggested that auxins, cytokinins, mesoinositol extra sucrose may have an important role in cambial development.

## Culture of Other Organs

Thanks to totipotentiality and regenerative ability of plant tissue in isolated conditions any living part or cell can be used for the regeneration of plants in vitro. Protocol for the culture of major organs is given above. Similarly, the other parts like nodes, internodes etc. are also cultured with minor modifications with respect to the media composition.

# Chapter - 9

# Production of Haploids *in vitro*

Haploid plants are characterized by possessing only a single set of chromosomes (gametophytic number of chromosomes i.e. n) in the sporophyte. This is in contrast to diploids which contain two sets (2n) of chromosomes. Haploid plants are of great significance for the production of homozygous lines (homozygous plants) and for the improvement of plants in plant breeding programmes. The process of apomixis or parthenogenesis (development of embryo from an unfertilized egg) is responsible for the spontaneous natural production of haploids. Many attempts were made, both by in vivo and in vitro methods to develop haploids. The success was much higher by in vitro techniques. The existence of haploids was discovered (as early as 1921) by Bergner in *Datura stramonium*. Plant breeders have been conducting extensive research to develop haploids. The Indian scientists Guha and Maheshwari (1964) reported the direct development of haploid embryos and plantlets from microspores of Datura innoxia by the cultures of excised anthers.

Basically there are two types of haploids. Monohaploids are the haploids that possess half the number of chromosomes from a diploid species e.g. maize, barley. Polyhaplods are the haploids possessing half the number of chromosomes from a polyploid species are regarded as polyhaploids e.g. wheat, potato. It may be noted that when the term haploid is generally used it applies to any plant originating from a sporophyte (2n) and containing half the number (n) of chromosomes.

**In vitro methids for haploid production:** In the plant biotechnology programmes haploid production is achieved by two methods.

**1. Androgenesis:** Haploid production occurs through anther or pollen culture, and they are referred to as androgenic haploids.

**2. Gynogenesis:** Ovary or ovule culture that results in the production of haploids, known as gynogenic haploids.

Androgenesis: In androgenesis, the male gametophyte (microspore or immature pollen) produces haploid plant. The basic principle is to stop the development of pollen cell into a gamete (sex cell) and force it to develop into a haploid plant. There are two approaches in androgenesis— anther culture and pollen (microspore) culture. Young plants, grown under optimal conditions of light, temperature and humidity, are suitable for androgenesis.

Gynogenesis: Haploid plants can be developed from ovary or ovule cultures. It is possible to trigger female gametophytes (megaspores) of angiosperms to develop into a sporophyte. The plants so produced are referred to as gynogenic haploids. Gynogenic haploids were first developed by San Noem (1976) from the ovary cultures of Hordeum vulgare. This technique was later applied for raising haploid plants of rice, wheat, maize, sunflower, sugar beet and tobacco.

## Anther and Pollen Culture

Anther culture is a technique by which the developing anthers at a precise and critical stage are excised aseptically from unopened flower bud and are cultured on a nutrient medium where the microspores within the cultured anther develop into callus tissue or embryoids that give rise to haploid plantlets either through organogenesis or embryogenesis. Pollen or microspore culture is an in vitro technique by which the pollen grains, preferably at the uni-nucleated stage, are squeezed out aseptically from the intact anther and then cultured on nutrient medium where the microspores, without producing male gametes, develop into haploid embryoids or callus tissue that give rise to haploid plantlets by embryogenesis or organogenesis.

### There are two modes of androgenesis

#### i) Direct androgenesis:

In this type, microspore behaves like a zygote and undergoes change to form embryoid which ultimately gives rise to a plantlet.

#### ii) Indirect androgenesis:

In contrast to the direct androgenesis, the microspore, instead of undergoing embryogenesis, divide, repeatedly to form a callus tissue which differentiates into haploid plantlets.

## Principle of Anther and Pollen Culture

The basic principle of anther and pollen culture is the production of haploid plants exploiting the totipotency of microspore and the occurrence of single set of chromosome (n) in microspore. In this process, the normal development and function of the pollen cell to become a male gamete is stopped and is diverted forcibly to a new metabolic pathway for vegetative cell division. For this objective, microspores, either within intact anther or in isolated state, are grown aseptically on the nutrient medium where the developing pollen grain will form the callus tissue or embryoids that ultimately give rise to haploid plantlets. In fact, anther culture is in essence the pollen culture. The principle behind the anther culture is that without disturbing the natural habitat and environment of the enclosed anther, pollen can be grown by culturing the intact anther. In culture condition, the diploid tissue of anther will remain living without proliferation at the selective medium and, at the same time, it will encourage the development of pollen by nursing and providing nutrient. The haploid embryoids or the callus tissue can be seen as the anther dehisces in culture But there is always the possibility that the diploid somatic cells of the anther will also respond to culture condition and so produce unwanted diploid callus or plantlets

In attempts to avoid this problem, free pollens isolated from the anther are grown in nutrient medium. The knowledge gained so far from anther and pollen culture has established that pollens at the uni-nucleate stage, just before the first mitosis, or during mitosis are most suitable for the induction of haploids. Induction of haploids can be enhanced by keeping the anther or flower bud at low temperature The low temperature has been ascribed to a number of factors such as dissolution of microtubules, alteration in the first mitosis or maintenance of higher ratio of viable pollen capable of embryogenesis. Cold treatment may also act to help the embryogenesis by repressing the gametophytic differentiation or by lowering the abscisic acid content of the anther which is considered to be inhibitory for the production of haploids. The important aspect of anther- or pollen culture is the nutrient medium. The nutritional requirements of the excised anther are much simpler than those of isolated microspores. In the isolated microspore, it is obvious that certain factors responsible for the induction of haploids, which might have been provided by the anther, are missing and these have to be provided through the medium.

Rich medium may encourage the proliferation of the diploid tissue of anther wall and should be avoided. Incorporation of activated charcoal into the medium has stimulated the induction of androgenesis. The iron in the medium also plays a very important role for the induction of haploids. Potato extract, coconut milk and growth regulators like auxin and cytokinin are also used for anther and pollen culture due to their stimulatory effect on androgenesis. In culture, pollen may divide mitotically or can follow the normal pathway of forming vegetative and generative nuclei. The generative nucleus remains quiescent and abort. The vegetative nucleus divides repeatedly, forming a multinucleate pollen. The multinucleate pollen undergoes segmentation which may lead to form either organised embryoid structure or callus tissue (Fig 11.1). Both types of development are utilised to form haploid plantlets.

The haploid plantlets are self-sterile due to presence of single set of chromosomes which are not able to participate in meiotic segregation. By colchicine treatment, haploids are made homozygous diploid, or isogeneic diploid which are fertile. Haploids or homozygous diploid grown.

### Comparison between anther and pollen cultures

Anther culture is easy, quick and practicable. Anther walls act as conditioning factors and promote culture growth. Thus, anther cultures are reasonably efficient for haploid production. The major limitation is that the plants not only originate from pollen but also from other parts of anther. This results in the population of plants at different ploidy levels (diploids, aneuploids). The disadvantages associated with anther culture can be overcome by pollen culture.

### Advantages of pollen culture

. Undesirable effects of anther wall and associated tissues can be avoided.

i. Androgenesis, starting from a single cell, can be better regulated.

ii. Isolated microspores (pollen) are ideal for various genetic manipulations (transformation, mutagenesis).

v. The yield of haploid plants is relatively higher.

**In vitro Pathways of pollen:** The cultured microspores mainly follow four distinct pathways during the initial stages of in vitro androgenesis.

**Pathway I:** The uninucleate microspore undergoes equal division to form two daughter cells of equal size e.g. *Datura innoxia*.

**Pathway II:** In certain plants, the microspore divides unequally to give bigger vegetative cell and a smaller generative cell. It is the vegetative cell that undergoes further divisions to form callus or embryo. The generative cell, on the other hand, degenerates after one or two divisions—e.g., *Nicotiana tabacum* and *Capsicum annuum*.

**Pathway III:** In this case, the microspore undergoes unequal division. The embryos are formed from the generative cell while the vegetative cell does not divide at all or undergoes limited number of divisions e.g. *HyoScyamus niger*.

**Pathway IV:** The microspore divides unequally as in pathways I and II. However, in this case, both vegetative and generative cells can further divide and contribute to the development of haploid plant e.g. *Datura metel* and *Atropa belladonna*.

At the initial stages, the microspore may follow any one of the four pathways described above. As the cells divide, the pollen grain becomes multicellular and burst open. This multicellular mass may form a callus which later differentiates into a plant (through callus phase). Alternately, the multicellular mass may produce the plant through direct embryogenesis (Fig. 45.1).

**Factors affecting the in vitro response:** A good knowledge of the various factors that influence androgenesis will help to improve the production of androgenic haploids. Some of these factors are briefly described. These factors are more or less same for gynogenesis also.

**Genotype of donar plants:** The success of anther or pollen culture largely depends on the genotype of the donor plant. It is therefore important to select only highly responsive genotypes. Some workers choose a breeding approach for improvement of genotype before they are used in androgenesis.

**Stage of microspore or pollen:** The selection of anthers at an ideal stage of microspore development is very critical for haploid production. In general, microspores ranging from tetrad to bi-nucleate stages are more responsive. Anthers at a very young stage (with microspore mother cells or tetrads) and late stage (with bi-nucleate microspores) are usually not suitable for androgenesis. However, for maximum production of androgenic haploids, the suitable stage of microspore development is dependent on the plant species, and has to be carefully selected.

**Physiological status of a donar plant:** The plants grown under best natural environmental conditions (light, temperature, nutrition, $CO_2$ etc.) with good anthers and healthy microspores are most suitable as donor plants. Flowers obtained from young plants, at the beginning of the flowering season are highly responsive. The use of pesticides should be avoided at least 3-4 weeks preceding sampling.

**Pretreatment of anthers:** The basic principle of native androgenesis is to stop the conversion of pollen cell into a gamete, and force its development into a plant. This is in fact an abnormal pathway induced to achieve in vitro androgenesis. Appropriate treatment of anthers is required for good success of haploid production.

**Treatment methods are variable and largely depend on the donor plant species:**

1.  **Chemical treatment:** Certain chemicals are known to induce parthenogenesis e.g. 2-chloroethylphosphonic acid (ethrel). When plants are treated with ethreal, multinucleated pollens are produced. These pollens when cultured may form embryos.

2.  **Temperature influence:** In general, when the buds are treated with cold temperatures (3-6°C) for about 3 days, induction occurs to yield pollen embryos in some plants e.g. Datura, Nicotiana. Further, induction of androgenesis is better if anthers are stored at low temperature, prior to culture e.g. maize, rye. There are also reports that pretreatment of anthers of certain plants at higher temperatures (35°C) stimulates androgenesis e.g. some species of Brassica and Capsicum.

**Effect of light:** In general, the production of haploids is better in light. There are however, certain plants which can grow well in both light and dark. Isolated pollen (not the anther) appears to be sensitive to light. Thus, low intensity of light promotes development of embryos in pollen cultures e.g. tobacco.

**Effect of culture medium:** The success of another culture and androgenesis is also dependent on the composition of the medium. There is, however, no single medium suitable for anther cultures of all plant species. The commonly used media for anther cultures are MS, White's, Nitsch and Nitsch, N6 and B5. These media in fact are the same as used in plant cell and tissue cultures. In recent years, some workers have developed specially designed media for anther cultures of cereals.

Sucrose, nitrate, ammonium salts, amino acids and minerals are essential for androgenesis. In some species, growth regulators — auxin and/or cytokinin are required for optimal growth. In certain plant species, addition of glutathione and ascorbic acid promotes androgenesis. When the anther culture medium is supplemented with activated charcoal, enhanced androgenesis is observed. It is believed that the activated charcoal removes the inhibitors from the medium and facilitates haploid formation.

## Production of Haploids by Ovary Culture

Ovary culture is a technique of culture of ovaries isolated either from pollinated or un-pollinated flowers.

**Principle:**

Ovary is a ovule bearing region of a pistil. Excised ovaries can be cultured in vitro. For many species e.g. tomato, gherkin (*Cucumis anguria*) excised ovaries grow in culture and form the fruits that ripen and produce viable seeds. This development takes place on a simple nutrient medium containing only mineral salts and sucrose, provided the flowers have been fertilized two or more days before excision.

But the ovaries of un-pollinated flowers do not grow on simple nutrient medium. However, use of some synthetic auxins such as 2, 4-D, 2 4 5- T (2, 4-5-trichlorophenoxyacetic acid), NOA (2, Napthoxyacetic acid) in the nutrient medium induces the development of ovaries of un-pollinated flowers.

Often, in culture, the ovaries fail to grow into full size fruits in the restricted space of culture vial. To overcome this problem, a partial sterile culture technique is devised in which only the long flower stalk is inserted into the aseptic nutrient medium through an opening in the stopper, thus leaving the ovary free to grow outside the culture vial.

**Protocol:**

1. Collect the pollinated or un-pollinated flowers from a healthy plant.

2. Wash them thoroughly with tap water, dip into 5% Teepol solution for 10 minutes and again wash to remove the trace of Teepol.

3. Transfer the flowers to laminar air flow cabinet. Surface sterilizes the flowers by immersing in 5% sodium hypochlorite solution for 5-7 minutes. Wash them with sterile distilled water.

4. Transfer the flowers to a sterile petri dish. Using a flamed force and a surgical scalpel, dissect out the calyx, petals, anther filaments etc. of the flower to isolate the pistil. During isolation of pistils care should be taken to ensure that the ovaries are not injured in any way. Damaged pistil, should be discarded as they often form callus tissue from the damaged parts.

5. Place the ovaries on agar solidified nutrient medium.

6. Incubate the cultures at 25°C in a 16 hrs, daylight regime at about 2000 lux The light is provided by fluorescent lamp.

**Importance:**

Ovary culture is a useful technique to investigate many fundamental and applied aspects.

Importance of Ovary Culture:

1. It is useful to study the early development of embryo development, fruit development, and different aspects of fruit physiology including respiration, maturation and disease.

2. The effect of phytohormones on parthenocarpic fruit development can be studied from the culture of un-pollinated pistil.

3. Floral organs play a significant role in fruit development. Role of floral organs can be studied from the in vitro culture of ovary. In some cases it has been found that pollinated ovary produce the normal fruits in vitro if the sepals are not removed before culture.

   However, if the sepals are removed before culture, addition of sucrose (5%) is necessary to obtain satisfactory growth of ovary in culture. In barley, lemma and palea are very important. In onion, the growth of ovaries without perianth is markedly inhibitory.

4. In hybridization, the plant breeders face many problems such as the failure of pollen germination on the stigma or the slow and insufficient growth of pollen tube as well as precocious abscission of flowers. Ovary or pistil culture, in vitro fertilization (test tube pollination) has been used to circumvent these obstacles. In many cases, successful results of in vitro fertilization and seed formation has been obtained.

   Petunia Violaceae is a self-incompatible species. But in vitro pollination and seed formation overcome the self-incompatibility. Actually, in vitro fertilization has got the immense value in plant breeding where problems of incompatibility or sterility exist as a barrier to normal sexual reproduction.

   It might be possible, therefore, to produce a hybrid between two species or varieties in vitro that could not be produced under normal in vivo conditions. Ovary culture has been successfully employed to obtain hybrids of diploid *Brassica chinensis* and autotetraploid *B. chinensis* which are normally cross incompatible.

5. Ovary culture has also been successful in inducing polyembryony. Poly-embryo

may develop in culture from the various parts of the ovary. These poly-embryos give rise to many shoots instead of a single plantlet.

6. The process of double fertilization not only brings about the formation of embryo and endosperm, but also stimulates the development of ovary into fruit. In most apomictic plants, although there is no fertilization, pollination alone stimulates the growth of the ovary and seed. The culture of ovaries of apomicts may, therefore, help in understanding the nature of stimulus provided by pollination.

## Ovule Culture

It is an elegant experimental system by which ovules are aseptically isolated from the ovary and are grown aseptically on chemically defined nutrient medium under controlled conditions. An ovule is a mega sporangium covered by integument. Ovules are attached with placenta inside the ovary by means of its funiculus. An ovule contains a megaspore or an egg cell. After fertilization, a single cell zygote is formed which ultimately leads to form a mature embryo possessing shoot and root primordia. Ovules can be isolated and cultured in nutrient medium. In vitro ovule culture helps to understand the factors that regulate the development of a zygote through organised stages to a mature embryo. Alternatively, it may be possible to germinate pollen in the same culture as the excised ovule and to induce in vitro fertilisation and subsequently embryo production.

### Protocol:

1. Collect the open flower (unfertilized ovules). If fertilized ovules are desired, collect the open flower where the anthers are dehisced and pollination has taken place. To ensure the fertilization, collect the flower after 48 hrs. of anther dehiscence

2. Remove sepals, petals, androecium etc. from the ovaries containing either fertilized or unfertilized ovules.

3. Soak the ovaries in 6% NaOCl solution.

4. Rinse the ovaries 3-4 times with sterile distilled water.

5. Using sterile techniques, ovules are gently prodded with the help of spoon shaped spatula by breaking the funicles at its junction with placental tissue.

6. The spatula with ovules is gently lowered into the sterile solid or liquid medium as the culture vial is slanted about 45°.

7. Damaged or undersized ovules are rejected when possible, during transfer.

8. Incubate the ovule culture in either dark or light (16 hrs. 3,000 lux) at 25°.C

## Importance of ovule culture:

Isolated ovule culture as early as the zygote or two to four celled pro-embryo stage is of considerable importance. Ovule culture is a boon for the plant breeders in obtaining seedlings from crosses which are normally unsuccessful because of abortive embryos.

## Limitations of ovule culture:

1. The dissection of unfertilized ovaries and ovules is rather difficult.

2. The presence of only one ovary per flower is another disadvantage. In contrast, there are a large number of microspores in one another. However, the future of gynogenesis may be more promising with improved and refined methods.

## Importance and application of in vitro ovule culture are discussed below on different specific aspects

### Test Tube Pollination and Fertilization:

An important achievement of research on ovule culture has been the development of the technique of test tube pollination and fertilization. By this technique, it may be possible to germinate pollen in the same culture as the excised ovule and to induce in vitro fertilization.

Excised unfertilized ovules of *Argemone mexicana, Eschscholtzia califormca, Papaver sonniferum, Nxcotxana tabacum* and *N. rustica* have been cultured along with their respective pollen grams. All the stages of development starting from the germination of pollen to double fertilization have been observed and the mature seeds containing viable embryos have been obtained by the above experiments.

Using the same method, it has been possible to fertilize the ovules of *Melandrium album* with pollen grains from other species of caryophyllaceae and subsequently even with pollen of *Datura stramonium*. Employing ovule culture technique, the incompatibility barrier in *Petunia axillaris* has been overcome.

### Application of Ovule Culture in Hybridization:

In many interspecific and inter-generic crosses, the $F_1$ hybrid embryos frequently become abortive in the developing seeds or the $F_1$ seeds are not capable to support the development of embryos. Ovule culture has been successfully employed to obtain hybrid seedlings. It has been observed that in several inter specific crosses; the hybrid embryo of *Abelmoschus* fails to develop beyond the heart or torpedo-shaped embryo.

By ovule culture, viable hybrids have been obtained in three out of five interspecific crosses attempted, namely, *A esculentus x A ficuneus A esculentus x A moschatus and A tuberculatus x A moschatus*. Similarly, a true hybrid *between Brassica* chinensis and *B. pekinensis* has been obtained by culturing the fertilized ovule in vitro. A hybrid between *Lolium perenne* and *Festuca rubra* has also been obtained successfully by means of ovule culture.

Several attempts have been made to hybridize between different species of the New World and Old World cotton. Although successful crosses between different species of cotton have been achieved, hybrid plants have not been obtained through fertilized ovule culture. But the seed development and the production of fibre from the cultured ovule have been demonstrated.

The in vitro growth ovule and the development of fibre from the develop- mg seed can also be regulated by exogenous hormones and in this respect ovule culture of cotton offers an unique method for the studies on the effect of phytohormones on fibre and seed development.

**Production of Haploid Callus through Ovule Culture:**

Uchimiya et al. (1971) attempted culturing unfertilized ovules of *Solanum melongena* and obtained vigorous callus formation on a medium supplemented with IAA and kinetin. Although the origin of the callus tissue was not known, a cytological assay revealed it to be haploid in nature. So it is an important attempt to obtain a haploid cell line or plant from an alternative source rather than anther or pollen culture.

**Ovule Culture and Angiospermic Parasites:**

It is generally believed that in obligate root parasites such as *Striga* or *Orobanke* the formation of seedlings is dependent on some stimulus from the host root. Studies on ovule culture of *Orobanche aegyptica* and *Cistanche tubulosa* have demonstrated that the formation of shoots in vitro can be induced in any absence of any stimulus from the host.

## Ovule Culture of Orchid Plants:

In nature, the seeds of orchid germinate only in association with a proper fungus. As a result numerous seeds are lost due to unavailability of proper fungus. Beside this, the seed capsule of many orchid takes a long time to mature. To overcome such problems, several attempts have been made to culture the fertilized ovule of orchid in vitro. Poddubnaya-Arnoldi (1959, '60) successfully grew the fertilized ovule of *Calanthe veitchn, Cypripedium insigne, Dendrobium no- bile* and *Phalaenopsis schilleriana.*

## Induction of Poly-embryo by Ovule Culture:

In horticultural practices, the artificial induction of poly-embryo holds a great potential. It has been observed that the nucellus of mono-embryonic ovule of citrus can be induced to form adventive embryos in culture. Therefore, such achievement is very significant.

## Virus Irradiation through Ovule Culture:

In the varieties of Citrus which are impossible to free of virus by other means, the ovule culture has proved decisively advantageous to make them virus free.

## Anther and Pollen Culture:

By careful selection of developing anthers at a precise and critical stage, it is possible to establish the anther culture that will give rise to haploid plantlets. Alternatively, the developing pollen grains can be diverted from their normal pathway and are induced to form somatic embryoids which subsequently give rise to the haploid plantlets.

Haploid plants contain a single set of chromosomes. The phenotype is the expression of single copy of genetic information as there is no masking of characters because of gene dominance. The use of anther and pollen culture offers an important tool to the conventional plant breeder as a way of obtaining haploid plants for selection of characters and inclusion in the breeding programme.

## Identification of Haploids:

Two approaches based on morphology and genetics are commonly used to detect or identify haploids.

**Morphological approach:** The vegetative and floral parts and the cell sizes of haploid plants are relatively reduced when compared to diploid plants. By this way haploids can be detected in a population of diploids. Morphological approach, however, is not as effective as genetic approach.

**Genetic approach:** Genetic markers are widely used for the specific identification of haploids. Several markers are in use.

    i. '$a_1$' marker for brown coloured aleurone.

    ii. 'A' marker for purple colour.

    iii. 'Lg' marker for ligule less character.

The above markers have been used for the development of haploids of maize. It may be noted that for the detection of androgenic haploids, the dominant gene marker should be present in the female plant.

**Diploidizatioim of Haploid Plants (Production of Homozygous Plants):** Haploid plants are obtained either by androgenesis or gynogenesis. These plants may grow up to a flowering stage, but viable gametes cannot be formed due to lack of one set of homologous chromosomes. Consequently, there is no seed formation. Haploids can be diploidized (by duplication of chromosomes) to produce homozygous plants. There are mainly two approaches for diploidization— colchicine treatment and endomitosis.

**Colchicine treatment:** Colchicine is very widely used for diploidization of homologous chromosomes. It acts as an inhibitor of spindle formation during mitosis and induces chromosome duplication. There are many ways of colchicine treatment to achieve diploidization for production of homozygous plants.

1.   When the plants are mature, colchicine in the form of a paste is applied to the axils of leaves. Now, the main axis is decapitated. This stimulates the axillary buds to grow into diploid and fertile branches.

2.   The young plantlets are directly treated with colchicine solution, washed thoroughly and replanted. This results in homozygous plants.

3.   The axillary buds can be repeatedly treated with colchicine cotton wool for about 2-3 weeks.

**Endomitosis:** Endomitosis is the phenomenon of doubling the number of chromosomes without division of the nucleus. The haploid cells, in general, are unstable in culture with a tendency to undergo endomitosis. This property of haploid cells is exploited for diploidization to produce homozygous plants.

The procedure involves growing a small segment of haploid plant stem in a suitable medium supplemented with growth regulators (auxin and cytokinin). This induces callus formation followed by differentiation. During the growth of callus, chromosomal doubling occurs by endomitosis. This results in the production of diploid homozygous cells and ultimately plants.

# Chapter - 10

# Single Cell Culture

Single cell culture is a method of growing isolated single cell aseptically on a nutrient medium under controlled condition. The basic principle of single cell culture is the isolation of large number of intact living cells and cultures them on a suitable nutrient medium for their requisite growth and development. Single cells can be isolated from a variety of tissue and organ of green plant as well as from callus tissue and cell suspension. Single cells from the intact plant tissue (leaf, stem, root cladode etc.) are isolated either mechanically or enzymatically. Mechanical isolation involves tearing or chopping the surface sterilized explant to expose the cells followed by scraping of the cells with a fine scalpel to liberate the single cells hoping that it remains undamaged. But very few living cells are obtained for a lot of time and effort. Gentle grinding of surface sterilized explant in a sterilized mortar-pestle followed by cleaning the cells by filtration and centrifugation is now widely used for the large-scale mechanical isolation of viable cells. A considerably more efficient way of large-scale isolation of free cells from the surface sterilized is to dissolve the intercellular cementing material, i.e. pectin, by pectinase or macerozyme treatment. The enzyme macerates the tissue from which large-number of variable cells can be obtained. The special feature of enzymatic isolation of cell is that it has been possible to obtain pure preparation of viable cells with less effort and time.

The isolated single cell can be cultured either in liquid medium or on solid medium. There are five basic methods that are used for culturing single cells such as paper raft nurse technique the petri dish plating technique, the micro-chamber technique, the micro-droplet technique, the plating with nurse tissue technique. In culture, the single cells divide re-divide to form a callus tissue. Such callus tissue also retains the capacity to regenerate the plantlets through organogenesis and embryogenesis.

## Applications of single cell cultures

PTC techniques are becoming increasingly popular as an alternative means of plant vegetative propagation, mass production of chemicals, and genetic engineering. The primary goal of plant tissue culture is crop management. This involves asexual methods of propagation to generate whole plants from single cells. In order for successful selection and genetic manipulation to occur, one must have successful *in vitro* plant regeneration, making this topic very important. *In vitro* clonal propagation is referred to as micropropagation. In this context clonal means the production of genetically identical plants grown from parts of plants and reproduced by asexual reproduction. This procedure allows for many thousands of plants to be derived from a single cell or tissue in a relatively short amount of time. This appears to be advantageous over conventional plant reproduction in that only a small amount of plant tissue is required as the initial explant for regeneration of millions of clonal plants in one year. Therefore, this *in vitro* technique allows for speedy international exchange of plant materials. It also eliminates the danger of introducing disease to the crop if performed under sterile conditions. Furthermore, when crops are grown *in vitro* it allows for year round growth regardless of what Mother Nature throws at us. Unfortunately, it is very expensive to carry out clonal propagation due to the costly equipment and trained engineers to perform the procedures. In addition to high growth rates there could also be high contamination rates. Once contamination occurs it could spread through the crops causing great losses over a very short amount of time. Cell cultures are also useful for the secondary metabolites they produce. Some of these metabolites that are a valuable source include flavors, natural sweeteners, industrial feed stocks, perfumes and commercial insecticides. These products do not perform vital physiological functions like amino acids or nucleic acids, but they are produced to ward off potential predators, attract pollinators, or combat infectious diseases. Another useful metabolite produced by plants includes shikokin, which is a chemical used as both a dye and a pharmaceutical. Plant cell suspension culture is valuable for studying the biosynthesis of secondary metabolites. Although there are limitations of cell culture systems in producing secondary metabolites, they are favored over conventional cultivation methods. This is because of their ability to produce useful compounds under controlled conditions as well as their capability of using this technique to produce chemicals to meet market demands. In addition, specific cells of a plant can be multiplied to produce a higher yield of specific metabolites, which cannot be done through conventional methods of cultivation.

## Techniques to Culture Plant Cell

### 1. The Paper Raft Nurse Technique:

i.   Single cells are isolated from suspension cultures or a friable callus with the help of a micropipette or micro-spatula.

ii.  Few days before cell isolation, sterile 8 mm x 8 mm squares of filter paper are placed aseptically on the upper surface of the actively growing callus tissue of the same or different species.

iii. The filter paper will be wetted by soaking the water and nutrient from the callus tissue.

iv.  The isolated single cell is placed aseptically on the wet filter paper raft.

v.   The whole culture system is incubated under 16 hrs. cool white light (3,000 lux) or under continuous darkness at 25° C.

vi.  The single cell divides and re-divides and ultimately forms a small cell colony. When the cell colony reaches a suitable size, it is transferred to fresh medium where it gives rise to the callus tissue.

The callus tissue, on which the single cell is growing, is called the nurse tissue. Actually the callus tissue supplies the cell with not only the nutrients from the culture medium but something more that is critical for cell division. The single cell absorbs nutrients through filter paper. The nutrients actually diffuse upward from culture medium through callus tissue and filter paper to the single cell. A callus tissue originating from a single cell is known as a single cell clone.

### 2. The Petri Dish Plating Technique:

i.   A suspension of purely single cells is prepared aseptically from the stock cell suspension culture by filtering and centrifugation requisite cell density in the single cell suspension is adjusted by adding or reducing the liquid medium.

ii.  The solid medium (1.6% 'Difco' agar added) is melted in water bath.

iii. In front of laminar air flow, the tight lid of falcon plastic petri dish is opened. With the help of sterilized Pasteur pipette 1 5 ml of single cell suspension is put an equal amount of melted agar medium when it cools down at 35°C, is added in the single cell suspension (Fig 9.2).

## 3. The Micro-chamber Technique:

i.   A drop of liquid nutrient medium containing single cell is first isolated aseptically from stock suspension culture with the help of long fine Pasteur pipette.

ii.  The culture drop is placed on the centre of a ' sterile microscopic slide (25 x 75 mm) and ringed with sterile paraffin oil.

iii. A drop of paraffin oil is placed on either side of the culture drop and a cover-glass (called raiser) is placed on each oil drop.

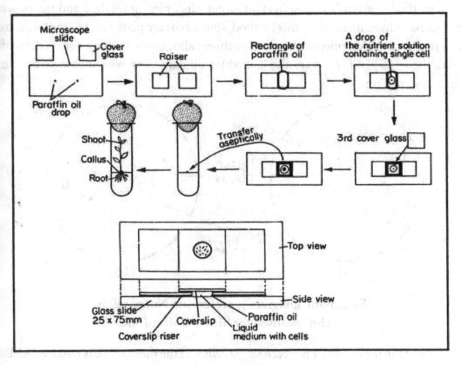

**Schematic Diagram representing Micro- Chamber technique used for Single Cell culture**

iv.  A third cover-glass is then placed on the 'culture drop bridging the two-raiser cover-glasses and forming a micro-chamber to enclose the single cell aseptically within the paraffin oil. The oil prevents the water loss from the culture drop but permits gaseous exchange.

v.   The whole micro-chamber slide is placed in a petri-dish and is incubated under 16 hrs. white cool illumination (3,000 lux) at 25 C.

vi.  Cell colony derived from the single cell gives rise to single cell clone.

vii.  When the cell colony becomes sufficiently large, the cover-glass is removed and the tissue is transferred to fresh solid or semisolid medium.

The micro-chamber technique permits regular observation of the growing and dividing cell.

### 4. The Nurse Callus Technique:

This method is actually a modification of petridish plating method and the paper raft nurse culture method. In this method, single cells are plated on to agar medium in a petridish as described earlier Two or three callus masses (Nurse tissue) derived from the same plant tissue are also embedded directly along with the single cells in the same medium.

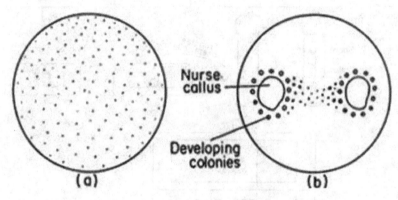

**Schematic Diagram representing Nurse callus
technique used for Single Cell culture**

Here the paper barrier between single cells and the nurse tissue is removed. Cells first begin to divide in the regions near the nurse callus indicating that the single cells closer to nurse callus in the solid medium gets the essential growth factors that are liberated from the callus mass. The developing colonies growing near to nurse callus also stimulate the division and colony formation of other cells.

### 5. The Micro-droplet Technique:

i.  In this method, single cells are cultured in special Cuprak dishes which have two chambers—a small outer chamber and a large inner chamber. The large chamber carries numerous numbered wells each with a capacity of 0.25-25µl of nutrient medium.

i. Each well of the inner chamber is filled with a micro-drop of liquid medium containing isolated single cell. The outer chamber is filled with sterile distilled water to maintain the humidity inside the dish.

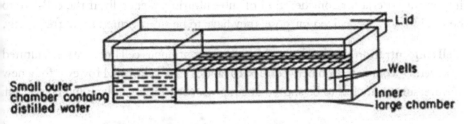

**Schematic Diagram representing Micro-droplet technique used for Single Cell culture**

ii. After covering the dish with lid, the dish is sealed with paraffin.

iv. The dish is incubated under 16hrs cool light (3,000 lux) at 25°C.

v. The cell colony derived from the single cell is transferred on to fresh solid or semisolid medium in a culture tube for further growth.

## Methods of Cell Dissociation

There are physical and enzymatic methods for dissociation of monolayer cultures. Mechanical shaking and cell scraping are employed for cultures which are loosely adhered, and the use of proteases has to be avoided. Among the enzymes, trypsin is the most frequently used. For certain cell monolayers, which cannot be dissociated by trypsin, other enzymes such as pronase, dispase and collagenase are used. Prior to cell dissociation by enzymes, the monolayers are usually subjected to pretreatment of EDTA for the removal of $Ca^{2+}$.

**Culture density:** It is advisable to subculture the normal or transformed monolayer cultures, as soon as they reach confluence. Confluence denotes the culture stage wherein all the available growth area is utilized and the cells make close contact with each other.

**Medium exhaustion:** A drop in pH is usually accompanied by an increase in culture cell density. Thus, when the pH falls, the medium must be changed, followed by subculture.

**Scheduled timings of subculture:** It is now possible to have specified schedule timings for subculture of each cell line. For a majority of cell cultures, the medium change is usually done after 3-4 days, and sub-culturing after 7 days.

**Purpose of subculture:** The purpose for which the cells are required is another important criteria for consideration of sub-culturing. Generally, if the cells are to be used for any specialized purpose, they have to be sub-cultured more frequently.

**Cell concentration at subculture:** Most of the continuous cell lines are sub-cultured at a seeding concentration between $1 \times 10^4$ and $5 \times 10^4$ cells/ ml. However for a new culture, subculture has to be started at a high concentration and gradually reduced.

# Chapter - 11

# Cell Suspension Cultures

Cell suspension is a suspension of cell in liquid medium. Suspension culture is a type of culture in which single cells or small aggregates of cells obtained from plant explants or callus multiply while suspended in agitated liquid medium. It is also referred to as cell culture or cell suspension culture. However single cell culture is different from cell suspension culture. Callus proliferates as an unorganised mass of cells. So, it is very difficult to follow many cellular events during its growth and developmental phases. To overcome such limitations of callus culture, the cultivation of free cells as well as small cell aggregates in a chemically defined liquid medium as a suspension was initiated to study the morphological and biochemical changes during their growth and developmental phases. To achieve an ideal cell suspension, most commonly a friable callus is transferred to agitated liquid medium where it breaks up and readily disperses. After eliminating the large callus pieces, only single cells and small cell aggregates are again transferred to fresh medium and after two or three weeks a suspension of actively growing cells is produced. This suspension can then be propagated by regular sub-culture of an aliquot to fresh medium. Ideally suspension culture should consist of only single cells which are physiologically and biochemically uniform. Although this ideal culture has yet to be achieved, but it can be achieved if it is possible to synchronize the process of cell division, enlargement and differentiation within the cell population.

The culture of single cells and cell aggregates of plant origin in moving liquid medium can be handled as the culture of microbes. The suspension culture eliminates many of the disadvantages ascribed to the callus culture on agar medium. Movement of the cells in relation to nutrient medium facilitates gaseous exchange, removes any polarity of the cells due to gravity and eliminates the nutrient gradients within the medium and at the surface of the cells.

Single culture is discussed in a separate chapter.

**Protocol:**

1. Take 150/250 ml conical flask containing autoclaved 40/60 ml liquid medium.

2. Transfer 3-4 pieces of pre-established callus tissue (approx. wt. 1 gm. each) from the culture tube using the spoon headed spatula to conical flasks.

3. Flame the neck of conical flask, close the mouth of the flask with a piece of alluminium foil or a cotton plug. Cover the closure with a piece of brown paper.

4. Place the flasks within the clamps of a rotary shaker moving at the 80-120 rpm (revolution per minute)

5. After 7 days, pour the contents of each flask through the sterilized sieve pore diameter -60µ- 100µ and collect the filtrate in a big sterilized container. The filtrate contains only free cells and cell aggregates.

6. Allow the filtrate to settle for 10-15 min. or centrifuge the filtrate at 500 to 1,000 rpm and finally pour off the supernatant.

7. Re-suspend the residue cells in a requisite volume of fresh liquid medium and dispense the cell suspension equally in several sterilized flasks (150/250 ml). Place the flasks on shaker and allow the free cells and cell aggregates to grow.

8. At the next subculture, repeat the previous steps but take only one-fifth of the residual cells as the inoculum and dispense equally in flasks and again place them on shaker.

9. After 3-4 subcultures, transfer 10 ml of cell suspension from each flask into new flask containing 30 ml fresh liquid medium.

10. To prepare a growth curve of cells in suspension, transfer a definite number of cells measured accurately by a haemocytometer to a definite volume of liquid medium and incubates on shaker. Pipette out very little aliquot of cell suspension at short intervals of time (1 or 2 days interval) and count the cell number. Plot the cell count data of a passage on a graph paper and the curve will indicate the growth pattern of suspension culture.

## Importance of Cell Suspension Culture

The culture of single cells and small aggregates in moving liquid medium is an important experimental technique for a lot of studies that are not correctly possible to do from the callus culture. Such a system is capable of contributing much significant information about cell physiology, biochemistry, metabolic events at the level of individual cells and small cell aggregates. It is also important to build up an understanding of an organ formation or embryoid formation starting from single cell or small cell aggregates. The technique of plating out cell suspension on agar plates is of particular value where attempts are being made to obtain single cell clones. Suspension culture derived from medicinally important plants can be studied for the production of secondary metabolites such as alkaloids and a considerable amount of industrial effort is being placed on the exploitation and expansion of this area. Mutagenesis studies may be facilitated by the use of cell suspension cultures to produce mutant cell clones from which mutant plants can be raised. Cell population in a suspension can be subjected to a range of mutagenic chemicals e.g. ethyl methane-sulphonate (EMS), N-nitroso-N- methyl urea etc. The mutagens can be added directly in the liquid medium. After the mutagen treatment, cells are plated on agar medium for the selection of mutant cell clones. The hope is that permanent changes in the DNA patterns of some of the cells would be achieved by such treatments. Plants could be raised from the mutant cell clones and the mutant plants are selected from the population either by morphological differences or by metabolic/biochemical differences. The selected plants can then be grown on and propagated further to produce a mutant population for evaluation studies.

**Types of cell suspension culture:** Batch culture and continuous culture are two types of techniques employed to cultivate microorganisms in large scale for industrial and other purposes. In batch culture, microorganisms are provided with nutrients at the beginning and grown. When microbes utilize the available nutrients, nutrients become limited after certain time period. Microorganisms grow via lag, log, stationery and death phases. The fermentation process is carried out batch-wise in batch culture technique. After each and every batch, fermenter is cleaned and used freshly for the next batch. In continuous culture, microorganisms are provided with adequate levels of fresh nutrients continually to always maintain the microbes at log phase to extract primary metabolites of the microorganisms. The volume of the continuous culture is maintained at a constant value by adding fresh nutrients and removing products at the same rate without stopping the process. Batch culture is needed comparatively a large closed fermenter while continuous culture is needed a small open fermenter. This is the difference between batch and continuous culture.

**Batch culture:** Batch culture is a technique which grows microorganisms in a closed system where a limited amount of nutrients are supplied at the beginning. This is the commonest technique adopted in industries to make useful products using microorganisms such as bacteria and fungi. Microbe which grows in the fermenter ferments the nutrients. Fermentation is a breaking down process of carbohydrates into alcohols and acids by microorganisms under anoxic conditions. In batch culture technique, nutrients are provided at the beginning and the particular microorganism is inoculated into the fermenter. The fermenter is closed and the temperature and pH are maintained for the growth of microorganisms. Microorganism grows inside and utilizes the provided nutrients and other conditions. With time, nutrients become limited and the environmental conditions change within the fermenter. Hence, microbial growth shows distinct four stages such as lag phase, log phase, stationary phase, and death phase. At the end of the fermentation, the process is stopped and useful products are extracted and purified. The fermenter is washed and sterilized before using for another batch culture.

The specialty of the batch culture technique is, this is run under limited amounts of nutrients and for a certain time period. The fermenter setup is easy to make and handle. Environmental conditions inside the fermenter vary with time. However, required temperature, pH conditions, stirring, pressure, etc. are properly maintained to achieve successful product formation.

Batch culture technique is widely used for the purification of secondary metabolites such as antibiotics, pigments, etc. This technique is not suitable for the production of primary metabolites and products which are associated with growth.

**Continuous culture:** Continuous culture is another technique which grows useful microorganisms. It aims to maintain a continuously growing microbial culture at exponential phase. It can be achieved by supplying fresh nutrients continually, removing accumulated waste and products at the same rate and keeping other conditions at the optimum values. It is done inside a special chamber called chemostat. Fresh medium is added continuously from one end while metabolic products are continually extracted from the other end of the chemostat to keep the culture volume at a constant level. Continuous culture is used in industries when it is required to extract useful primary metabolites such as amino acids, organic acids, etc. from the microorganisms. Primary metabolites are produced at the highest rate when the microorganisms are at their exponential phase. Hence continuous culture always aims to maintain the microbial biomass at the log phase. It is done by monitoring the process continually and controlling the system.

**Types of batch culture:** Batch culture is a type of suspension culture where the cell material grows in a finite volume of agitated liquid medium. For instance, cell material in 20 ml or 40 ml or 60 ml liquid medium in each passage constitute a batch culture. Batch suspension cultures are most commonly maintained in conical flasks incubated on orbital platform shakers at the speed of 80-120 rpm. The following are the different types of batch culture.

**Slowly rotating cultures:** Single cells and cell aggregates are grown in a specially designed flask, the nipple flask. Each nipple flask possesses eight nipple-like projections. The capacity of each flask is 250 ml. Ten flasks are loaded in a circular manner on the large flat disc of a vertical shaker. When the flat disc rotates at the speed of 1-2 rpm, the cell within each nipple of the flask are alternately bathed in culture medium and exposed to air.

**Shake cultures:** It is a very simple and effective system of suspension culture. In this method, single cells and cell aggregates in fixed volume of liquid medium are placed in conical flasks. Conical flasks are mounted with the help of clips on a horizontal large square plate of an orbital platform shaker. The square plate moves by a circular motion at the speed of 60-180 rpm.

**Spinning cultures:** Large volumes of cell suspension may be cultured in 10L bottles which are rotated in a culture spinner at 120 rpm at an angle of 45°.

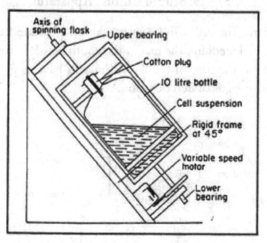

**Diagram of a 10 litre spinning culture apparatus**

**Stirred culture:** This system is also used for large-scale batch culture (1.5 to 10 litres). In this method, the large culture vessel is not rotated but the cell suspension inside the vessel is kept dispersed continuously by bubbling sterile air through culture medium. The use of an internal magnetic stirrer is the most convenient way to agitate the culture medium safely. The magnetic stirrer revolves at 200-600 rpm. The culture vessel is a 5 to 10 litres round-bottom flask.

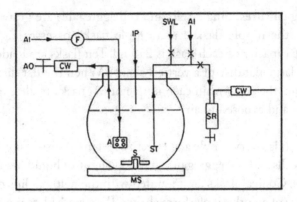

Stirred batch culture unit. Arrow indicate direction of flow of air; AI = air input; F = sterilizing glass-fibre air filter; AO = air outlet; CW = cotton wool; IP = inoculation port; A = aerator; S = stirrer magnet; ST = sample tube; MS = magnetic stirrer; SWL = sterile water line; SR = sample receiver; (Diagram after Dr. P. King)

### Fig.19: Stirred Culture Apparatus

**Stirrer flask culture:** The cells can be suspended in a culture flask (a stirrer flask) containing the desired medium. The medium is continuously stirred with a magnetic pendulum rotating at the base of the flask. The cells have to periodically examine for contamination or signs of deterioration.

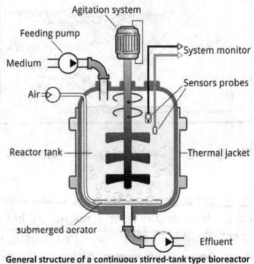

General structure of a continuous stirred-tank type bioreactor

### Stirred Tank Bioreactor (Image Courtesy: G. Yassine Mrabet CC BY-SA 4.0, https://commons.wikimedia.org)

**Continuous culture:** This system is very much similar to stirred culture. But in this system, the old liquid medium is continuously replaced by the fresh liquid medium to stabilize the physiological states of the growing cells. Normally, the liquid medium is not changed until the depletion of some nutrients in the medium and the cells are kept in the same medium for a certain period. As a result active growth phase of the cell declines the depletion of nutrient. In continuous culture system, nutrient depletion does not occur due to continuous flow of nutrient medium and the cells always remain in the steady state of active growth phase.

**Chemostats:** In this system, culture vessels are generally cylindrical or circular in shape and possess inlet and outlet pores for aeration and the introduction of and removal of cells and medium. The liquid medium containing the cells is stirred by a magnetic stirrer. The introduction of fresh sterile medium, which is pumped in at a constant rate into the vessel is balanced by the displacement of an equal volume of spent or old medium and cells. Such a system can be maintained in a steady state so that new cells are produced by division at a rate which compensates the number lost in the outflow of spent medium. Thus in a steady state condition the density, growth rate, chemical composition and metabolic activity of the cells all remain constant. Such continuous cultures are ideal for studying growth kinetics and the regulation of metabolic activity in higher plants.

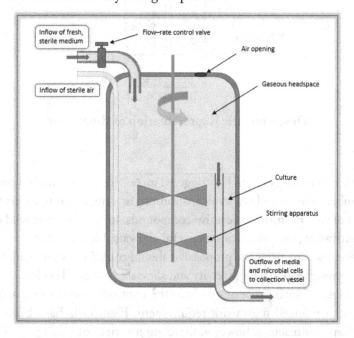

**Diagrammatic Representation of Bioreactor (Image Courtesy: CGraham2332**

**Turbidostats:** The turbidity of a suspension culture medium changes rapidly when the cells increase in number due to their steady state growth. The changes of turbidity of the culture medium can be measured by the changes of optical density of the medium. Again, the pH of the medium changes due to increase of cell density. In turbidostates, an automatic monitoring unit is connected with the culture vessel and such unit adjusts the medium flow in such a way as to maintain the optical density or pH at a chosen, preset level.

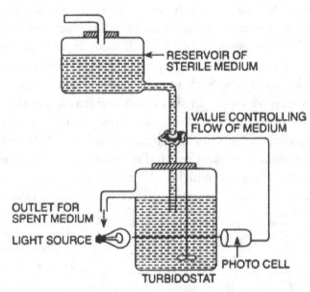

**Diagrammatic Representation of Bioreactor**

## Bioreactors

The recent biotechnology boom has triggered increased interest in plant cell cultures since a number of firms and academic institutions investigated intensively to rise the production of very promising bioactive compounds. In alternative to wild collection or plant cultivation, the production of useful and valuable secondary metabolites in large bioreactors is an attractive proposal; it should contribute significantly to future attempts to preserve global biodiversity and alleviate associated ecological problems. The advantages of such processes include the controlled production according to demand and a reduced man work requirement. Plant cells have been grown in different shape bioreactors, however, there are a variety of problems to be solved before this technology can be adopted on a wide scale for the production of useful plant secondary metabolites. There are different factors affecting the culture growth and secondary metabolite production in bioreactors: the gaseous atmosphere, oxygen supply and $CO_2$ exchange, pH, minerals, carbohydrates, growth regulators, the liquid medium rheology and cell density. Moreover agitation systems and sterilization conditions may negatively influence the whole process. Many types of bioreactors have been successfully used for cultivating transformed root cultures, depending on both different aeration system and nutrient supply.

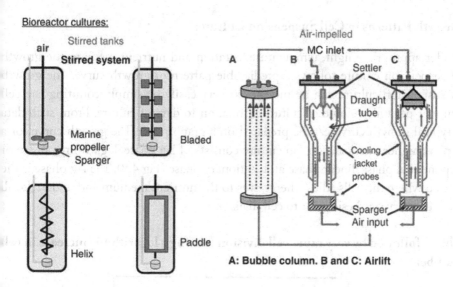

A: Bubble column. B and C: Airlift

## Advantages of cell suspension culture:

. The process of propagation is faster.

i. The lag period is usually shorter.

ii. Results in homogeneous suspension of cells.

v. Treatment with trypsin is not required.

. Scale-up is convenient.

i. No need for frequent replacement of the medium.

ii. Maintenance is easy.

iii. Bulk production of cells can be conveniently achieved.

## Criteria for Cell Suspension Subculture:

The criteria adopted for suspension subculture are the same as that already described for monolayer subcultures.

**The following aspects have to be considered:**

i. Culture density.

ii. pH change representing medium exhaustion.

iii. Schedule timings of subculture.

iv. Purpose of subculture.

**Growth Patterns in Cell Suspension Culture:**

Under appropriate light, temperature, aeration and nutrient medium the growth of suspension culture follows a predictable pattern or growth curve. The growth of suspension culture can be monitored very easily by simply counting the cell number per unit volume of culture in relation to days of culture. From such data a typical growth curve can be prepared on a graph paper. The growth curve for a typical higher plant suspension culture consists of lag phase, logarithmic phase or exponential phase, linear phase and stationary phase (Fig 4.8). The lag phase is the period where the cells adjust themselves to the nutrient medium and undertake all the necessary synthesis prior to cell division.

**This is followed by very rapid cell division causing a logarithmic increase in cell number.**

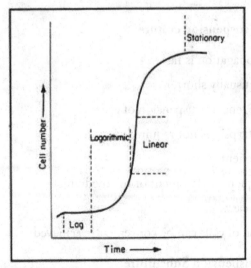

**Growth Curve: Graphical representation of the Growth**

This phase is called as logarithmic phase. A further period of rapid cell division results in a linear increase in number and the phase is called linear phase. As nutrients are depleted and some of the cells of the culture being to show senescent characteristics, the rate of cell division within the culture declines and it passes through the stationary phase. At this stage the growth curve forms a plateau. If the cells are removed just before or just after the entry into stationary phase in each growth cycle and are sub-cultured to fresh medium, then identical patterns of growth of the cell line can be maintained in each culture passage. Dry weight, total protein, DNA synthesis etc. can also be considered as other parameters for the preparation of identical growth curves. It also indicates that the chemical composition of the cell changes throughout the growth cycle and such changes are closely coupled to the cell division in most of the plant material. However, in some material there is no corresponding increase in dry weight accumulation and consequently the divergence between the rate of cell division and rate of dry weight accumulation increases. From these studies, it has been concluded that there are independent mechanisms for controlling cell division and many biosynthetic pathways.

A synchronized cell population and the continuous changes in physiological property may also cause the divergence between the rate of cell division and the biochemical changes of the cell. It is also important to note that the degree of cellular aggregation is not constant but changes significantly during the growth cycle of the suspension culture. As the culture enters the period of most active growth the cell aggregation is maximum and during the stationary phase cell aggregation is minimal. For experimental studies on growth of cell suspension, the inoculum or cell density is an important factor. Very low density or high density of cells in liquid medium is unable to grow. So, to induce the growth, an initial density of $2 \times 10^6$ cells/ml to $2 \times 10^8$ cells/ml is inoculated in liquid medium. This initial density increases during growth and attains a higher density at the stationary phase. Most commonly, such high cellular density are diluted on subculture by a factor of ca. X 10. The particular initial cell density that is able to grow in liquid medium is called critical initial density (CID). The CID may vary from plant to plant.

## Measurement of Cell Growth in Suspension Culture

The cells in suspension culture grow by cell division and the number of cells increases. Growth studies of this kind are very valuable for the characterization of cell lines, effect of nutrient medium and hormones etc. Growth in such cultures can be monitored by determination of cell number, cell dry weight, packed cell volume, etc. The rate of increase of cell number can be calculated simply by counting the cell number in haemocytometer under a microscope. Cell count- data obtained from haemocytometer is multiplied by a factor x $10^3$ and the result can be expressed in terms of cell number per unit volume of culture. Therefore, by comparing the cell number at the beginning of culture and after certain days of incubation, the growth can be measured. Again, as the cells increase in number during growth, the liquid medium will be more turbid and as a result the optical density (OD) of the suspension culture will also be altered. The changes of OD value can be detected by a calorimeter. Therefore, from OD value growth can be measured. Definite volumes of cell suspension can be harvested from multiple replicated sets of culture. Such amount of cell suspension is transferred in a graduated conical centrifuge tube and is centrifuged at 2,000X g for 5 minutes. The cells will form a pellet after centrifugation. The volume of cell pellet then represents the packed cell volume (PCV). It is also called biomass volume.

Therefore, harvesting the cell suspension at definite periods of interval and measuring the PCV, the growth can be monitored and expressed as milliliter cell pellet per milliliter culture. From the same experiments dry weight of cell mass can also be estimated by drying the pellet in a hot air oven (12 hours at 60°C) after replacing the supernatant and weighing the dried cell mass in a chemical or electrical balance. In this method, growth can be expressed in terms of dry weight in gram or milligram per unit volume of culture.

## Test for Viability of Cell

Cell death may occur in suspension cultures due to several factors. So, for the studies on growth the test for viability of cells is very important. Otherwise, cell count data will be erroneous without testing the viability. The most frequently used staining method for assessing cell viability is fluorescein diacetate (FDA). FDA dissolved in 5 mg/ml of acetone is added to cell population at 0.01% final concentration. Dead cells fluoresces red. Evans blue also used at a final concentration of 0.01% is specific for dead cells. As soon as the stain is mixed with cell suspension, the in- viable cells stain blue and the viable cells remain unstained.

# Differences between Batch and Continuous Culture

| Parameters | Batch Culture | Continuous Culture |
| --- | --- | --- |
| Definition | Batch culture technique is used to cultivate beneficial microorganisms under limited amounts of nutrients in a closed fermenter for a certain time period. Microbial growth inside the batch culture shows a typical microbial growth curve in which four distinct phases can be identified. | Continuous culture technique is used to grow beneficial microorganisms under optimum level of nutrients in an open system in which nutrients are added continually and waste and products are removed at the same rate to keep the growth at an exponential phase. |
| Nutrients | Nutrients are supplied once before starting the fermentation process. | Nutrients are added many times (at starting and in between the process). |
| Type of System | Batch culture is a closed system | Continuous culture is an open system. |
| Termination process | The process of the batch culture is stopped after the product is formed. | The process is not stopped though the product is formed. Continuous removal of the product is done without stopping the process in continuous culture. |
| Environmental conditions | The environmental conditions inside the batch culture are not constant. | The environmental conditions inside the continuous culture are maintained at constant level. |
| Cell growth | Microbial growth inside the batch culture follows lag, log and stationary phases. | Microbial growth is maintained at optimum level which is an exponential growth stage. |
| Turnover rate | Turnover rate is low since the nutrients and other conditions are limited inside. | Turnover rate is high since the optimum levels of nutrients and other conditions are maintained. |
| Culture setup | Batch culture setup is easy to make and run. | Continuous culture setup is not easy to make and run. |
| Contamination | Contaminations are minimum in batch cultures | Contamination chance is high in continuous culture. |
| Control methods | Control methods are easy and quick. | Control methods are complicated and time-consuming. |

| Suitability | Batch culture is more suitable for the production of secondary metabolites such as antibiotics. | Continuous culture is more suitable for the production of primary metabolites such as amino acids and organic acids. |
|---|---|---|
| Uses | Batch culture is commonly used in small scale product synthesis and academics | Continuous culture is less used in large scale industrial production |

# Chapter - 12

# Principles, Techniques
## of Plant Protoplast Culture

## What is a Protoplast?

It is known that each and every plant cell possesses a definite cellulosic cell wall and the protoplast lies within the cell wall except some reproductive cells and the free floating cells in some fruit juices like coconut water. Therefore, protoplast of plant cell consists of plasma-lemma and everything contained within it. But those of importance to plant protoplast culture are produced experimentally by the removal of cell wall by either enzymatically or mechanical means from the artificially plasmolysed plant cells. Experimentally produced protoplasts are known as isolated protoplasts. According to Torrey and Landgren (1977) "the isolated protoplasts are the cells with their walls stripped off and removed from the proximity of their neighbouring cells". Vasil (1980) defines that "the protoplast is a part of plant cell which lies within the cell wall and can be plasmolysed and which can be isolated by removing the cell wall by mechanical or enzymatic procedure". Therefore, isolated protoplast is only a naked plant cell surrounded by plasma membrane—which is potentially capable of cell wall regeneration, cell division, growth and plant regeneration in culture.

## Brief Past History

**J. Klercker (1892):** First isolated protoplast mechanically from plasmolyzed cell of water warrior (Stratiotes aloides). No attempt was made to culture them.

**E. Kiister (1927):** In the fruits of several plants like Solatium nigram, Lycopersicon esculenium etc. the cell wall are hydrolysed during fruits ripening process so that free protoplasts and protoplasmic units are left. Kuster preferred such physiological method for isolating protoplasts. No report of culture was available.

**R. Chambers and K. Hofler (1931):** Were able to isolate few protoplasts by using thin slices of epidermis of onion bulb immersed in 1M sucrose until the protoplast shrunk away from their enclosing walls and then cutting sheets of epidermis with a sharp knife. Report of culture was not available.

**E. C. Cocking (1960):** First reported the enzymatic method for isolation of protoplast in a large number from root tip cells of Lycopersicon esculentum by using a concentrated solution of cellulase enzyme, prepared from cultures of the fungus Myrothecium verrucaria to degrade cell wall.

**I. Takebe, Y. Otsuki and S. Aoki (1968):** First employed the commercial preparation of cellulase and macerozyme sequentially (in two steps) for the isolation of mesophyll protoplast of tobacco.

**J. B. Power and E. C. Cocking (1968):** Demonstrated first that the mixture of such two enzymes (cellulase + macerozyme) can be used simultaneously (one step method) for the isolation of protoplasts.

**I. Takebe, G. Labib, G. Melchers (1971):** First reported the plant regeneration from isolated protoplast in Nicotiana tabacum.

**P. S. Carlson, H. H. Smith, R. D. Dearing (1972):** First reported a somatic hybrid in higher plants involving two different sexually compatible species of mesophyll protoplast (N. glauca x N. langsdorffi).

**Different Sources of Plant Tissue and Their Condition for Protoplast Isolation:**

Protoplast can be isolated either directly from the different parts of whole plant or indirectly from in vitro cultured tissue. Convenient and suitable materials are leaf mesophyll and cells from liquid suspension cultures. Protoplast yield and viability are profoundly influenced by the growing conditions of plants serving leaf mesophyll sources. The age of the plant and of the leaf and the prevailing conditions of light, photoperiod, humidity, temperature, nutrition and watering are contributing factors. Cell suspension cultures may provide a more reliable source for obtaining consistent quality protoplasts. It is necessary, however, to establish and maintain the cells at maximum growth rates and utilize the cell at the early log phase.

**Principles of Protoplast Culture:**

The basic principle of protoplast culture is the aseptic isolation of large number of intact living protoplasts removing their cell wall and cultures them on a suitable nutrient medium for their requisite growth and development. Protoplast can be isolated from varieties of plant tissues. Convenient and suitable materials are leaf mesophyll and cells from liquid suspension culture. Protoplast yield and viability are greatly influenced by the growing condition of the plant as well as the cells. The essential step of the isolation of protoplast is the removal of the cell wall without damaging the cell or protoplasts. The plant cell is an osmotic system. The cell wall exerts the inward pressure upon the enclosed protoplasts. Likewise, the protoplast also puts equal and opposite pressure upon the cell wall. Thus, both the pressures are balanced. Now if the cell wall is removed, the balanced pressures will be disturbed. As a result, the outward pressure of protoplast will be greater and at the same time in absence of cell wall, irresistible expansion of protoplast takes place due to huge inflow of water from the external medium. Greater outward pressure and the expansion of protoplast cause it to burst. So, the isolated protoplast is an osmotically fragile structure at its nascent stage. Therefore, if the cell wall is to be removed to isolate protoplast, the cell or tissue must be placed in a hypertonic solution of a metabolically inert sugar such as mannitol at higher concentration (13%) to plasmolysis the cell away from the cell wall (Fig 12.1).

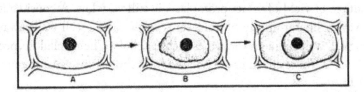

**Stages of Plasmolysis**

Mannitol, an alcoholic sugar, is easily transported across the plasmodesmata, provides a stable osmotic environment for the protoplasts and prevents the usual expansion and bursting of protoplast even after loss of cell wall. That is why, this hypertonic solution is known as osmotic stabilizer or plasmolyticum or osmolyticum. Once the cells are stabilized in such a manner by plasmolysis the protoplasts are released from the containing cell wall either mechanically or enzymatically. Mechanical isolation (Fig 12.2) involves breaking open each cell compartment to liberate the protoplast. This operation can be done carefully on small pieces of tissue under a microscope using a micro-scalpel.

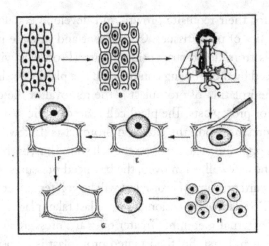

**Mechanical Isolation of Protoplasts**

But very few protoplasts are obtained for a lot of time and effort. Large-scale attempts at mechanical isolation involves the disrupting tissue with fine stainless steel-bristled brush. This process may liberate more protoplasts with less efforts, but the percentage of yield of intact protoplasts is still very low. A considerably more efficient way of liberating the protoplasts is to digest the cell walls away around them, using cell wall degrading enzymes such as cellulase, hemicellulose, pectinase or macerozyme etc. These enzymes are isolated from fungi and available commercially (Table 12.1).

Table 12.1  Commercial enzymes, their commercial name and source

| Enzyme | Source organism |
|---|---|
| **A. Cellulose degrading enzymes** | *Aspergillus niger* |
| Cellulysin (Onozuka R10) | *Trichoderma reessei (formally T. viride)* |
| Driselase | *Irpex lactes* |
| **B. Hemicellulose degrading enzymes** | |
| Hemicellulase | *Aspergillus niger* |
| Rhozyme HP150 | *A. niger* |
| **C. Pectin degrading enzymes** | |
| Pectinase | *A. niger* |
| Macerase (Macerozyme) | *Rhizopus spp.* |
| Pectinol AC, Pectolyase Y23 | *A. niger* |
| Pectic-acid acetyl transferase (PATE) | *A. japonicus* |

**List and source of enzymes that are used in Protoplast Isolation**

Period of treatment and concentration of enzymes are the critical factors and both factors should be standardized for particular plant tissue. Intact tissue can be incubated with a pectinase or macerozyme solution which will dissolve the middle lamella between the cells and so separate them.

Subsequent treatment with cellulase will digest away the cellulosic layer of the cell wall. This process is known as sequential enzyme treatment or two step method as opposed to a mixed enzyme treatment (one step method) in which both cellulase and pectinase or macerozyme are mixed so that the entire wall is broken down in a single operation.

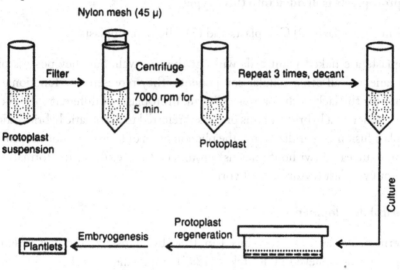

**Enzymatic Isolation of Protoplasts**

The isolated protoplasts can be cultured either static liquid or agarified medium. The protoplast media consist of mineral salts, vitamins, carbon sources and plant growth hormones as well as osmotic stabilizers and possibly organic nitrogen sources, coconut milk and organic acids.

In culture protoplast can reform a new cell wall around them. Once the wall is formed, the protoplast becomes a cell. The cells from protoplasts subsequently enter cell division which is followed by the formation of callus and cell cultures. Such callus also retain the capacity for morphogenesis and plant regeneration. A brief list of plant regeneration from plant protoplast culture is given below.

| Common name | Species | Family | Cell origin |
|---|---|---|---|
| Tobacco | *Nicotiana tabacum* | Solanaceae | Leaf, cell culture |
| Potato | *Solanum tuberosum* | Solanaceae | Leaf |
| Datura | *Datura innozia* | Solanaceae | Leaf |
| Petunia | *Petunia hybrida* | Solanaceae | Leaf |
| Carrot | *Daucus carota* | Umbelliferae | Cell culture |
| Rape seed | *Brassica napus* | Cruciferae | Leaf |
| Orange | *Citus sinensis* | Rutaceae | Nucellus callus |
| Asparagus | *Asparagus officinalis* | Liliaceae | Cladodes |
| Bromegrass | *Bromus inermis* | Poaceae | Cell culture |

## Sub-protoplasts is divided into three types

(1) Mini-protoplasts (2) Cytoplasts and (3) Micro-protoplasts.

Protoplasts are naked plant cells without the cell wall, but they possess plasma membrane and all other cellular components. They represent the functional plant cells but for the lack of the barrier, cell wall. Protoplasts of different species can be fused to generate a hybrid and this process is referred to as somatic hybridization (or protoplast fusion). Cybridization is the phenomenon of fusion of a normal protoplast with an enucleated (without nucleus) protoplast that results in the formation of a cybrid or cytoplast (cytoplasmic hybrids).

## Historical developments:

The term protoplast was introduced in 1880 by Hanstein. The first isolation of protoplasts was achieved by Klercker (1892) employing a mechanical method. A real beginning in protoplast research was made in 1960 by Cocking who used an enzymatic method for the removal of cell wall. Rakabe and his associates (1971) were successful to achieve the regeneration of whole tobacco plant from protoplasts. Rapid progress occurred after 1980 in protoplast fusion to improve plant genetic material, and the development of transgenic plants.

## Importance of Protoplasts and Their Cultures:

The isolation, culture and fusion of protoplasts is a fascinating field in plant research. Protoplast isolation and their cultures provide millions of single cells (comparable to microbial cells) for a variety of studies.

## Protoplasts have a wide range of applications; some of them are listed below:

1.  The protoplast in culture can be regenerated into a whole plant.

2. Hybrids can be developed from protoplast fusion.

3. It is easy to perform single cell cloning with protoplasts.

4. Genetic transformations can be achieved through genetic engineering of protoplast DNA.

5. Protoplasts are excellent materials for ultra-structural studies.

6. Isolation of cell organelles and chromosomes is easy from protoplasts.

7. Protoplasts are useful for membrane studies (transport and uptake processes).

8. Isolation of mutants from protoplast cultures is easy.

## Isolation of Protoplasts:

Protoplasts are isolated by two techniques

1. Mechanical method

2. Enzymatic method

## Mechanical method:

Protoplas isolation by mechanical method is a crude and tedious procedure. This results in the isolation of a very small number of protoplasts.

## The technique involves the following stages

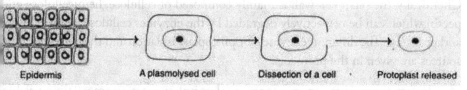

| Epidermis | A plasmolysed cell | Dissection of a cell | Protoplast released |

**Mechanical Isolation of Protoplasts**

1. A small piece of epidermis from a plant is selected.

2. The cells are subjected to plasmolysis. This causes protoplasts to shrink away from the cell walls.

3. The tissue is dissected to release the protoplasts.

**Mechanical method for protoplast isolation is no more in use because of the following limitations:**

i. Yield of protoplasts and their viability is low.
ii. It is restricted to certain tissues with vacuolated cells.
iii. The method is laborious and tedious.

However, some workers prefer mechanical methods if the cell wall degrading enzymes (of enzymatic method) cause deleterious effects to protoplasts.

**Enzymatic method:**

Enzymatic method is a very widely used technique for the isolation of protoplasts. The advantages of enzymatic method include good yield of viable cells, and minimal or no damage to the protoplasts.

**Sources of protoplasts:**

Protoplasts can be isolated from a wide variety of tissues and organs that include leaves, roots, shoot apices, fruits, embryos and microspores. Among these, the mesophyll tissue of fully expanded leaves of young plants or new shoots are most frequently used. In addition, callus and suspension cultures also serve as good sources for protoplast isolation.

**Enzymes for protoplast isolation:**

The enzymes that can digest the cell walls are required for protoplast isolation. Chemically, the plant cell wall is mainly composed of cellulose, hemicellulose and pectin which can be respectively degraded by the enzymes cellulose, hemicellulose and pectinase. The different enzymes for protoplast isolation and the corresponding sources are given in the protocols.

In fact, the various enzymes for protoplast isolation are commercially available. The enzymes are usually used at a pH 4.5 to 6.0, temperature 25-30°C with a wide variation in incubation period that may range from half an hour to 20 hours.

**The enzymatic isolation of protoplasts can be carried out by two approaches:**

**1. Two step or sequential method:** The tissue is first treated with pectinase (macerozyme) to separate cells by degrading middle lamella. These free cells are then exposed to cellulose to release protoplasts. Pectinase breaks up the cell aggregates into individual cells while cellulose removes the cell wall proper.

**2. One step or simultaneous method:** This is the preferred method for protoplast isolation. It involves the simultaneous use of both the enzymes — macerozyme and cellulose.

## Isolation of protoplasts from leaves:

Leaves are most commonly used, for protoplast isolation, since it is possible to isolate uniform cells in large numbers.

1. Sterilization of leaves.
2. Removal of epidermal cell layer.
3. Treatment with enzymes.
4. Isolation of protoplasts.

Besides leaves, callus cultures and cell suspension cultures can also be used for the isolation of protoplasts. For this purpose, young and actively growing cells are preferred.

**The procedure broadly involves the following steps**

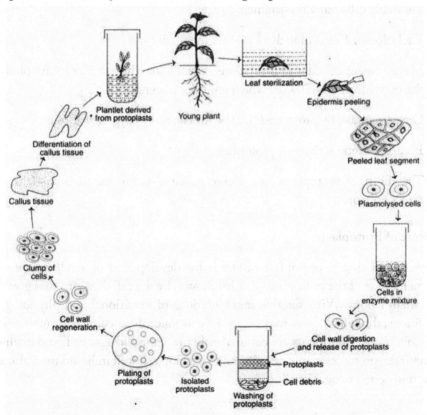

**Purification of protoplasts:**

The enzyme digested plant cells, besides protoplasts contain undigested cells, broken protoplasts and undigested tissues. The cell clumps and undigested tissues can be removed by filtration. This is followed by centrifugation and washings of the protoplasts. After centrifugation, the protoplasts are recovered above Percoll.

**Viability of protoplasts:**

It is essential to ensure that the isolated protoplasts are healthy and viable so that they are capable of undergoing sustained cell divisions and regeneration.

**There are several methods to assess the protoplast viability:**

1. Fluorescein diacetate (FDA) staining method—The dye accumulates inside viable protoplasts which can be detected by fluorescence microscopy.

2. Phenosafranine stain is selectively taken up by dead protoplasts (turn red) while the viable cells remain unstained.

3. Exclusion of Evans blue dye by intact membranes.

4. Measurement of cell wall formation—Calcofluor white (CFW) stain binds to the newly formed cell walls which emit fluorescence.

5. Oxygen uptake by protoplasts can be measured by oxygen electrode.

6. Photosynthetic activity of protoplasts.

7. The ability of protoplasts to undergo continuous mitotic divisions (this is a direct measure).

**Culture of Protoplasts:**

The very first step in protoplast culture is the development of a cell wall around the membrane of the protoplast. This is followed by the cell divisions that give rise to a small colony. With suitable manipulations of nutritional and physiological conditions, the cell colonies may be grown continuously as cultures or regenerated to whole plants. Protoplasts are cultured either in semisolid agar or liquid medium. Sometimes, protoplasts are first allowed to develop cell wall in liquid medium, and then transferred to agar medium.

## Agar culture:

Agarose is the most frequently used agar to solidify the culture media. The concentration of the agar should be such that it forms a soft agar gel when mixed with the protoplast suspension. The plating of protoplasts is carried out by Bergmann's cell plating technique. In agar cultures, the protoplasts remain in a fixed position, divide and form cell clones. The advantage with agar culture is that clumping of protoplasts is avoided.

## Liquid Culture

**Liquid culture is the preferred method for protoplast cultivation for the following reasons:**

1. It is easy to dilute and transfer.

2. Density of the cells can be manipulated as desired.

3. For some plant species, the cells cannot divide in agar medium, therefore liquid medium is the only choice.

4. Osmotic pressure of liquid medium can be altered as desired.

## Culture media:

The culture media with regard to nutritional components and osmoticum are briefly described.

## Nutritional components:

In general, the nutritional requirements of protoplasts are similar to those of cultured plant cells (callus and suspension cultures). Mostly, MS and B5 media with suitable modifications are used.

**Some of the special features of protoplast culture media are listed below:**

1. The medium should be devoid of ammonium, and the quantities of iron and zinc should be less.

2. The concentration of calcium should be 2-4-times higher than used for cell cultures. This is needed for membrane stability.

3. High auxin/kinetin ratio is suitable to induce cell divisions while high kinetin/auxin ratio is required for regeneration.

4.  Glucose is the preferred carbon source by protoplasts although a combination of sugars (glucose and sucrose) can be used.

5.  The vitamins used for protoplast cultures are the same as used in standard tissue culture media.

**Osmoticum and osmotic pressure:**

Osmoticum broadly refers to the reagents/ chemicals that are added to increase the osmotic pressure of a liquid.

The isolation and culture of protoplasts require osmotic protection until they develop a strong cell wall. In fact, if the freshly isolated protoplasts are directly added to the normal culture medium, they will burst.

Thus, addition of an osmoticum is essential for both isolation and culture media of protoplast to prevent their rupture. The osmotica are of two types — non-ionic and ionic.

**Non-ionic osmotica:**

The non-ionic substances most commonly used are soluble carbohydrates such as mannitol, sorbitol, glucose, fructose, galactose and sucrose. Mannitol, being metabolically inert, is most frequently used.

**Ionic osmotica:**

Potassium chloride, calcium chloride and magnesium phosphate are the ionic substances in use to maintain osmotic pressure. When the protoplasts are transferred to a culture medium, the use of metabolically active osmotic stabilizers (e.g., glucose, sucrose) along with metabolically inert osmotic stabilizers (mannitol) is advantageous. As the growth of protoplasts and cell wall regeneration occurs, the metabolically active compounds are utilized, and this results in the reduced osmotic pressure so that proper osmolarity is maintained.

**Culture Methods**

The culture techniques of protoplasts are almost the same that are used for cell culture with suitable modifications. Some important aspects are briefly given.

## Feeder layer technique:

For culture of protoplasts at low density feeder layer technique is preferred. This method is also important for selection of specific mutant or hybrid cells on plates. The technique consists of exposing protoplast cell suspensions to X-rays (to inhibit cell division with good metabolic activity) and then plating them on agar plates.

## Co-culture of protoplasts:

Protoplasts of two different plant species (one slow growing and another fast growing) can be co- cultured. This type of culture is advantageous since the growing species provide the growth factors and other chemicals which help in the generation of cell wall and cell division. The co-culture method is generally used if the two types of protoplasts are morphologically distinct.

## Micro drop culture:

Specially designed dishes namely cuprak dishes with outer and inner chambers are used for micro drop culture. The inner chamber carries several wells wherein the individual protoplasts in droplets of nutrient medium can be added. The outer chamber is filled with water to maintain humidity. This method allows the culture of fewer protoplasts for droplet of the medium.

## Regeneration of Protoplasts:

**Protoplast regeneration which may also be regarded as protoplast development occurs in two stages:**

1. Formation of cell wall.

2. Development of callus/whole plant.

**Formation of cell wall:** The process of cell wall formation in cultured protoplasts starts within a few hours after isolation that may take two to several days under suitable conditions. As the cell wall development occurs, the protoplasts lose their characteristic spherical shape. The newly developed cell wall by protoplasts can be identified by using calcofluor white fluorescent stain. The freshly formed cell wall is composed of loosely bound micro fibrils which get organized to form a typical cell wall. This process of cell wall development requires continuous supply of nutrients, particularly a readily metabolised carbon source (e.g. sucrose). Cell wall development is found to be improper in the presence of ionic osmotic stabilizers in the medium.

The protoplasts with proper cell wall development undergo normal cell division. On the other hand, protoplasts with poorly regenerated cell wall show budding and fail to undergo normal mitosis.

**Development of Callus/whole Plant:** As the cell wall formation around protoplasts is complete, the cells increase in size, and the first division generally occurs within 2-7 days. Subsequent divisions result in small colonies, and by the end of third week, visible colonies (macroscopic colonies) are formed. These colonies are then transferred to an osmotic-free (mannitol or sorbitol-free) medium for further development to form callus. With induction and appropriate manipulations, the callus can undergo organogenic or embryo genic differentiation to finally form the whole plant. A general view of the protoplast isolation, culture and regeneration is represented in Fig. 44.2. Plant regeneration can be done from the callus obtained either from protoplasts or from the culture of plant organs. There are however, certain differences in these two calluses. The callus derived from plant organs carries preformed buds or organized structures, while the callus from protoplast culture does not have such structures. The first success of regeneration of plants from protoplast cultures of Nicotiana tabacum was achieved by Takebe et al (in 1971). Since then, several species of plants have been regenerated by using protoplasts.

### Protocol for isolation and culture of protoplast

Protoplasts can be prepared from a variety of tissue but among them leaf mesophyll tissue from a wide range of plants has been proved to be the most ideal source of plant material for protoplast isolation. Leaves of Nicotiana tabacum is a highly standardized material for easy entry into the art of protoplast isolation and culture. Now-a-days the mixed enzyme method (single step) is very popular and this procedure is followed as routine work in most of the laboratories of the world.

**The protocol for isolation of protoplast from the mesophyll cells of tobacco using mixed enzyme method and its culture are described below:**

### Isolation of Protoplast

1. Young fully expanded leaves from the upper part of 7-8 weeks old plants growing in a greenhouse are detached and leaves are washed thoroughly with tap water

2. Surface sterilization of leaves is done by first immersing in 70% ethanol for 60 seconds followed by dipping into 0.4-0.5% sodium hypochlorite solution for 30 minutes. For this purpose, a sterile casserole dish is used as a sterilizing

container. Sterilization is done in front of laminar air flow.

3.  After 30 minutes, the sterilant is poured off and leaves are washed aseptically 3-4 times with autoclaved distilled water to remove every trace of hypochlorite.

4.  With the help of long sterilized forceps (8 inches), one leaf is transferred on a sterilized floor tile. The impermeable lower epidermis of the surface sterilized leaves is peeled off as completely as possible. During this process, a sterilized fine jeweler's force is inserted into a junction of the midrib and a lateral vein and the epidermis, is carefully peeled away at an angle to the main axis of the leaf.

**Note:** Where peeling of the leaf is not possible, slicing of the leaf into thin strips may be sufficient to allow entry of the enzymes through the cut edges of the strip.

5.  Peeled leaf pieces are placed lower surface down onto 30 ml sterilized CPW 13M*solution in a 14 cm petridish. When the liquid surface is completely covered with peeled leaf pieces, then the CPW 13M solution is pipetted off from the beneath of leaf pieces. The CPW 13M solution is replaced by bacterial filter sterilized solution of enzyme containing 2% cellulase (Onozuka R1O), 0.5% macerozyme in 13% manitol added CPW (pH 5.5).

6.  Leaf pieces in enzyme solution are incubated in the dark at 24-26°C for 16-18 hrs.

7.  Without disturbing the digested leaf pieces the enzyme solution is gently replaced by CPW 13M. Then digested leaf pieces are gently agitated and squeezed with sterile fine forceps to facilitate the release of the protoplast. The protoplast suspension is then allowed to pass through a 60 $\mu$-80$\mu$ nylon mesh to remove the larger pieces of undigested tissue.

8.  The filtrate is transferred to screw-capped centrifuge tube and is spinned for 5 minutes at 100 g.

9.  The protoplasts form the pellet. The supernatant is pipetted off and the pellet is re- suspended in CPW 21S solution. It is again centrifuged for 5-7 minutes at 200 g. The viable protoplasts will float at the top surface of CPW 21S in the form of dark green band while the remaining cells and debris will sink at the bottom of the tube.

10. The viable protoplasts are collected from the surface and are re-suspended again in CPW 13M to remove the sucrose. Centrifugations are repeated two-three times for washing.

11. Finally the protoplasts are suspended in measured volume of liquid Nagata and Tabeke medium (1971) supplemented with NAA (3 mg/L), 6-BAP (1 mg/L) and 13% mannitol.

---

* CPW means cell and protoplast washing medium. The compositiion of CPW is given below—

| | | |
|---|---|---|
| $KH_2PO_4$ | ... | 27.2 mg/L |
| $KNO_3$ | ... | 101 mg/L |
| $CaCl_2, 2H_2O$ | ... | 1480 mg/L |
| $MgSO_4, 7H_2O$ | ... | 246 mg/L |
| KI | ... | 0.16 mg/L |
| $CuSO_4, 5H_2O$ | ... | 0.025 mg/L |
| pH | ... | 5.8 |

CPW 13M = CPW + 13% mannitol (13M)
CPW 21S = CPW + 21% sucrose (21S)

---

**Isolation of Protoplast from Cell Suspension Culture:**

Rapidly growing cell suspension cultures are the most suitable material for protoplast isolation. A new cell suspension does not yield many protoplast until it has been sub-cultured at least twice. For protoplast isolation, suspension cultures are generally harvested at its early exponential growth phase or log phase. Older suspension cultures have a tendency to form elongated giant cells with thick wall. So it is very difficult to isolate the protoplast from such culture. Again, the presence of large number of cell aggregates in suspension culture is not desirable for the isolation of protoplast. Addition of colchicine and some chelating chemicals in suspension culture generally prevents the formation of cell aggregates. Sometimes very low concentration of cellulase (0.1%) is added in cell suspension culture two days before use to discourage the formation of thick wall.

**Step 1:**

**Filtering of cell suspension:**

The harvested cell suspension is passed through a coarse nylon sieve so that filtrate contains single cells as well as very small cell clumps.

## Step 2:

**Preparation of liquid medium free cells for plasmolysis:** The filtrate is allowed to settle out of the medium. Most of the medium is decanted off and the cells are transferred (by pouring) to a flask. Using the Pasteur pipette, all of the culture medium is removed. This is best achieved by drawing off medium from the base of the cell layer.

## Step 3:

**Pre-plasmolysis:** The cells are suspended in CPW13M solution for 1 hr. After 1 hr., the plasmolyticum is pipetted off.

## Step 4:

**Enzyme incubation:** The enzyme solution is added to the cells. The flask containing cells and enzyme are placed on the platform of a slowly rotating gyratory shaker (ca 30-40 rpm) for standardized period (4-6 hrs.).

## Step 5:

**Washing and purification:** Protoplast suspension is filtered through 60/i-100/i stainless steel sieve to remove the larger debris. The filtrate is transferred into centrifuge tubes and is spinned at 80 g for 5-10 minutes so as to sediment the protoplasts and then supernatant is pipetted off. The pellet is re-suspended in CPW 21S solution.

The protoplast suspension is again spinned at 100 g for 5-7 minutes. Viable and debris free protoplasts are collected at the surface. The protoplasts are washed with CPW 13M by repeated centrifugation and finally protoplasts are re-suspended in measured volume of liquid culture medium. Finally, the yield of protoplast is measured by counting in haemocytometer.

## Sample Protocol—Isolation of Protoplast from Cell Suspension Culture of *Daucus carota*:

1. Allow the cells to settle out of the medium and remove the supernatant with Pasteur pipette.
2. Add 20-30 ml of enzyme solution (Onozuka R10 cellulase – 2% and hemicellulose – 1% in 10% mannitol added CPW solution, pH 5.5) and incubate the cells at room temperature on the platform of slowly rotating gyratory shaker 30 rpm) for 4 hrs.

3. Filter protoplast suspension through 67 µ stainless steel sieve and transfer the filtrate in centrifuged tube. Spin at 80 g for 5 min.

4. Discard in supernatant and re-suspend in CPW 21S. Spin 100 g for 5-7 minutes.

5. Collect the protoplast from the surface of CPW 21S solution.

6. Wash the protoplast with CPW 13M solution by repeated centrifugation (3-4 times).

7. Finally transfer the protoplast to a measured volume of liquid $B_5$ medium supplemented with 2, 4-D(1 mg/L), kinetin (2 mg/L) and casein hydrolysate (1 gm./1).

**Isolation of Haploid Protoplast:**

Pollen mother cell (PMC), pollen tetrad (PT) and mature pollen are the natural source of haploid cells. Since they are produced in large number in anther and is very easy to squeeze out from anther, they are suitable material for the isolation of protoplast. Egg cell is also haploid cell, but generally single egg cell is produced per ovule and is very difficult to separate them from complex tissue integration. Isolation and yield of protoplast from haploid male cells depends upon the composition of cell wall. The cell wall of PMC and PT is made up of callose (un-branched 1-3 glucan) whereas the exine of mature pollen is coated with sporopollenin which is very difficult to digest by enzymes. The freshly isolated protoplast from mature pollen tends to fuse spontaneously to form multinucleate giant protoplast. Therefore, isolation and culture of protoplast have yielded only limited success. Better preparation of protoplasts is obtained from pollen mother cell or pollen tetrad than the mature pollen. The isolated protoplast from haploid cell is known haploid protoplast or gametoplast. Isolation and culture of haploid protoplast are useful for the induction mutation and to study the effect of irradiation as well as chemical mutagens.

**Isolation of Protoplast from PMC and PT:**

To ascertain the presence of PMC and PT in anther in relation to flower bud size the anthers are smeared with 2% acetocarmine before using them. Like leaf mesophyll tissue and cell suspension culture, the isolation of haploid protoplast is also multistep process. Approximately, the anticipated size of anther containing PMC and PT are aseptically removed with fine forceps. The content of this anther are squeezed out with the help of glass spatula. The pollen mother cell and the pollen tetrad come out as a milky fluid. It is treated with 0.75-1% helicase enzyme in 8-10% sucrose for about 30-45 minutes. Helicase is obtained from snail gut. Sometimes Zymolase an enzyme isolated from Arthobactor luteus is used for the isolation of protoplast from pollen tetrad. After incubation enzyme is carefully replaced by 10% sucrose and the protoplasts are allowed to settle. Protoplasts are rinsed by conventional washing medium. Finally, protoplasts are cultured in either liquid or solid medium.

## Culture of Protoplast:

Isolated protoplast can be cultured in several ways of which agar embedding technique in small petridish is commonly followed. In this technique, protoplast suspension is mixed with equal volume of melted 1.6% 'Difco' agarified medium (37° C) and the protoplast-agar mixture are poured into small petridish. In petridishes, embedding of protoplasts in solid agar medium is known as plating of protoplasts. The plated protoplast can be handled very easily and the agar medium provides a good support to the protoplast. In situ developmental stages of embedded protoplast can be studied under compound microscope. Besides this, separated clones derived from individual protoplast can be monitored.

## The method is described below:

1. The protoplasts in liquid NT medium* are counted with the help of haemocytometer. The protoplast deficit is adjusted to $1 \times 10^5$ to $2 \times 10^5$ protoplast/ml.

2. Agar solidified (1.6% 'Difco' agar) NT medium is melted.

3. The tight lid of Falcon plastic petridish (35 mm diameter 5 mm thickness) is opened and 1.5 ml of protoplast suspension is taken. To this equal aliquot of melted agar medium is added when it cools down at 37°C to 40°C.

4. The lid is quickly replaced tightly and the whole dish is swirled gently to disperse the protoplast-agar medium mixture uniformly throughout the dish.

5. The medium is allowed to solidify. The petridish is then inverted.

6. The culture is incubated at 25° C with 500 lux illumination (16 hrs. light) initially.

7. The cultures are sub-cultured periodically in the same solid medium (0.8% agar) with gradually reducing mannitol.

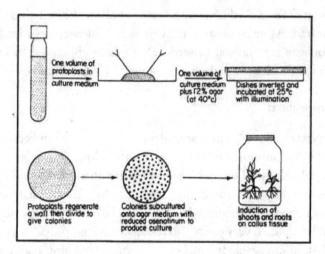

One volume of protoplasts in culture medium

One volume of culture medium plus 12% agar (at 40°c)

Dishes inverted and incubated at 25°c with illumination

Protoplasts regenerate a wall then divide to give colonies

Colonies subcultured onto agar medium with reduced osenotinum to produce culture

Induction of shoots and roots on callus tissue

**Schematic representation of Protoplast Culture**

Several other methods have been described for the culture of protoplasts, such as droplet culture, Co-culture, feeder layer, hanging droplets and immobilized/bead culture.

\* Composition of Nagata and Takebe (NT) medium (1971) for protoplast culture.

| Constituents | Amounts in mg/L |
|---|---|
| **Macro-nutrients** | |
| $NH_4NO_3$ | 825 |
| $KNO_3$ | 950 |
| $CaCl_2, 2H_2O$ | 220 |
| $MgSO_4, 7H_2O$ | 1233 |
| $KH_2PO_4$ | 680 |
| **Micro-nutrients** | |
| KI | 0.83 |
| $H_3BO_3$ | 6.2 |
| $MnSO_4, 4H_2O$ | 22.3 |
| $ZnSO_4, 4H_2O$ | 8.6 |
| $Na_2NoO_4, 5H_2O$ | 0.25 |
| $CuSO_4, 5H_2O$ | 0.025 |
| **Iron source** | |
| $FeSO_4, 7H_2O$ | 27.8 |
| $Na_2EDTA, 2H_2O$ | 37.3 |
| **Vitamins** | |
| Meso-inositol | 100 |
| Thiamine HCl | 1 |
| **Carbohydrate source** | |
| Sucrose | 1% |
| **Growth substances** | |
| α-napthalene-acid (NAA) | 3 |
| 6-Benzylaminopurine (6-BAP) | 1 |
| **Plasmolyticum** | |
| Mannitol | 13%–0% |
| pH | 5.8 |

For solid medium 1.6% or 0.8% agar is added

## Droplet Culture:

Suspending protoplasts in liquid culture media are placed on petri dishes in the form of droplet (Fig 12.7) with the help of micropipette. This method enables the subsequent microscopic examination of protoplast development. In this method, cultured protoplast clump together at the centre of droplets.

**Co-culture:** Sometimes a reliable fast growing protoplast is mixed in varying ratio with the less fast growing protoplast. The mixed protoplasts are plated in solid medium as described previously. The fast growing protoplast presumably provides some growth factors which induces the growth and development of the desirable protoplasts. This is known as co-culture technique.

## Feeder Layer Technique:

Fast growing protoplasts are sometimes made initotically blocked protoplast by low doses (1-2 Krad) of X-ray treatment. Such irradiated protoplasts are plated with agar medium. Upon this thin solidified layer of irradiated protoplast, desirable protoplasts are again plated at a low density with agar medium.

As a result, it makes two agar layers containing irradiated protoplast in lower layer and desirable protoplast in upper layer. The lower irradiated protoplast is known as feeder layer which improves the growth and development of normal protoplasts even at lower density.

## Hanging Droplet Method:

Culture of protoplasts in an inverted liquid droplet (0.25-0.50 µl) is known as hanging droplet method. Each droplet contains very small group of protoplasts. A number of droplets are generally placed on the inner surface of the lid of a petri dish.

Very thin layer of water is generally kept on the lower part of the petri dish to make a humid condition inside the petri dish as well as to prevent the dessication of the droplets. This technique facilitates to observe the development of protoplast under microscope. Protoplasts also gets better aeration as they go down to the hanging surface of the droplets.

## Bead Culture:

Sometimes protoplast suspension are mixed with several polymer like alginate, carrageenan etc. as well as melted standard difco agar. Small beads are made by dripping the mixture into liquid medium. After that, beads in liquid medium are put on moving shaker.

Entrapped protoplasts culture have shown several advantage over static liquid culture or slowly moving liquid culture where the protoplast suffers the mechanical breakage. This technique increases the mechanical stability, aeration and viability with biochemical activity.

**Tests for Viability of Protoplast:**

Viability of protoplasts after isolation and during culture in liquid medium is very important. Cell wall formation, cell division, callus formation etc. depend upon the viability of protoplast. The most frequently used staining methods for assessing protoplast viability are fluorescein diacetate (FDA), phenosafranine. FDA dissolved in 5.0 mg/ml acetone is added to the protoplast culture at 0.01% final concentration. The chlorophyll from broken protoplasts fluoresces red. Therefore, the percentage of viable protoplasts in a preparation can be easily calculated.

Phenosafranin, also used at a final concentration of 0.01% is specific for dead protoplast. As soon as the stain is mixed with protoplast preparation, the inviable protoplasts stain red and viable protoplasts remain unstained.

**Wall Formation, Cell Division and Callus Formation:**

The viable protoplast in culture regenerates its own wall around them. Once the wall is formed, the protoplast becomes essentially a regenerated cell. Depending upon the species, the protoplast remains in naked condition hardly for 10 minutes or a day. Generally protoplast begins to deposit cellulose micro-fibril immediately after washing and enzyme removal.

Cell membrane of newly isolated protoplast contains protruding microtubules that function in the orientation of newly synthesized cellulose micro-fibrils. The rate and regularity of cell wall regeneration depend on the plant species and the state of differentiation of the donor cell used for protoplast isolation. Calcafluor white (CFW) is the most commonly used stain to detect the onset of cell wall regeneration. CFW binds to the—linked glucosides in the newly synthesized cell wall. Optimum staining is achieved when 0.1 ml of protoplasts is mixed with 5.0 pi of a 0.1% v/v solution CFW. Cell wall synthesis is observed by a ring of white fluorescence around the plasma membrane. Cell wall regeneration is prerequisite for nuclear and cell division After the formation of cell wall, the walled cells expand and divide into two cells'

At this stage it looks like '8'. Cell division stages can also be stained using CFW. In most of cases, first cell division usually takes place with 2-7 days of culture. After the first division, each daughter cell divides into two cells. Repeated division results the formation of cell clump or cell aggregates. All the cells derived from the protoplasts do not divide and form the cell colonies. Therefore, the percentage of cells which give rises cell colonies, to known as plating (colony forming) efficiency. Several factors such as genotype of the donor plant, culture medium, hormones as well as physical factors are important for the division of protoplast and callus formation. The small callus mass can be handled in the conventional manner which means that it can be sub cultured at regular interval and can be used for organogenesis. For sub-culturing, the plate containing dividing protoplasts are sliced into several agar blocks. Each block is transferred to the surface of fresh medium. The plasmolyticum level in the culture medium is progressively reduced to zero by repeated sub-culturing.

## Plant Regeneration:

The ultimate objective in protoplast culture is the reconstruction of plant from the single protoplast. The strategy for plant regeneration has been to recover rapidly growing callus from protoplasts and to transfer the callus to a species specific regeneration medium (Fig 12.8). It is generally noted that plant regeneration occurs very easily in some plant species while others are recalcitrant. Plant regeneration from protoplast derived callus tissue have been reported mainly from solanaceous species. It includes 17 Nicotiana species, 6 Petunia species and 6 Solanum species. In recent years the list of non-solanaceous species capable of plant regeneration from protoplasts has been steadily expanding. The list contains several species mono- cot and dicot including carrot, endive, cassava, alfalfa, millet, clover, rapeseed, asparagus, cabbage, citrus etc.

## Sub-Protoplasts:

The fragments derived from protoplasts that do not contain all the contents of plant cells are referred to as sub-protoplasts. It is possible to experimentally induce fragmentation of protoplasts to form sub-protoplasts. This can be done by application of different centrifugal forces created by discontinuous gradients during centrifugation. Exposure of protoplasts to cytochalasin B in association with centrifugation is a better approach for fragmentation of protoplasts.

**There are three types of sub-protoplasts**

1. **Mini-protoplasts:** These are also called as karyoplasts and contain the nucleus. Mini-protoplasts can divide and are capable of regeneration into plants.

2. **Cytoplasts:** These are sub-protoplasts containing the original cytoplasmic material (in part or full) but lack nucleus. Thus, cytoplasts are nuclear-free sub-protoplasts which cannot divide, but they can be used for cybridization.

3. **Micro-protoplasts:** This term was suggested for sub-protoplasts that contain not all but a few chromosomes.

# Chapter - 13

# Agrobacterium Mediated Biotransformation

Plants provide human beings with all manner of useful products: food and animal feed, fibers and structural materials, and small molecules that can be used as dyes, scents, and medicines. People have sought to improve plants by breeding and selecting the better-performing and most useful varieties. The one limitation of this approach is that breeders are restricted to the existing gene pool in each species or sexually compatible group of species. Transgenic plants generated by direct DNA transfer methods (e.g., polyethylene glycol or liposome-mediated transformation, electroporation, or particle bombardment) often integrate a large number of copies of the transgene in tandem or inverted repeat arrays, in either multiple or single loci.

## Methods of delivering DNA into plant cells

Biological Mmethods;

- » Agrobacterium
- » Other bacteria
- » Viruses

Physical methods;

- » Particle bombardment
- » Electroporation
- » Silicon carbide whiskers
- » Carbon nanofibers

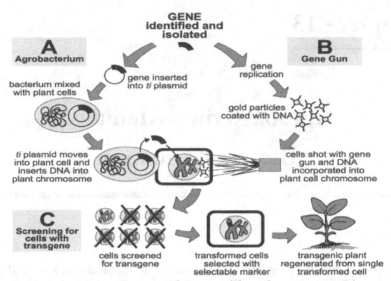

**Schematic representation of Genetic Transformation in Plants**

Twenty-five years ago, the concept of using *Agrobacterium tumefaciens* (soil gram-ve bacterium) as a vector to create transgenic plants (natural transformation) was viewed as a prospect and a "wish." Today, many agronomically and horticulturally important species are routinely transformed using this bacterium, and the list of species that is susceptible to *Agrobacterium*-mediated transformation seems to grow daily. In some developed countries, a high percentage of the acreage of such economically important crops as corn, soybeans, cotton, canola, potatoes, and tomatoes is transgenic; an increasing number of these transgenic varieties are or will soon be generated by Agrobacterium-mediated, as opposed to particle bombardment-mediated transformation. There still remain, however, many challenges for genotype-independent transformation of many economically important crop species, as well as forest species used for lumber, paper, and pulp production. In addition, predictable and stable expression of transgenes remains problematic. *A. tumefaciens* involvement in Crown gall disease was viewed by Smith & Townsend (1907). The stable transmission through the germ line was first demonstrated in 1981, when transgenic tobacco plants were generated by transformation using *A. tumefacians*.

### *Agrobacterium* "Species" and Host Range

The genus *Agrobacterium* has been divided into a number of species. However, this division has reflected, for the most part, disease symptomology and host range. Thus, *A. radiobacter* is an "avirulent" species, *A. tumefaciens* causes crown gall disease, *A. rhizogenes* causes hairy root disease, and *A. rubi* causes cane gall disease. More recently a new species has been proposed, *A. vitis*, which causes galls on grape and a few other plant species. We now know that symptoms follow, for the most part, the type of tumorigenic plasmid contained within a particular strain. Curing a particular plasmid and replacing this plasmid with another type of tumorigenic plasmid can alter disease symptoms. For example, infection of plants with *A. tumefaciens* C58, containing the nopaline-type Ti plasmid pTiC58, results in the formation of crown gall teratomas. When this plasmid is cured, the strain becomes nonpathogenic. Introduction of Ri plasmids into the cured strain "converts" the bacterium into a rhizogenic strain. Regardless of the current confusion in species classification, for the purposes of plant genetic engineering, the most important aspect may be the host range of different *Agrobacterium* strains. As a genus, *Agrobacterium* can transfer DNA to a remarkably broad group of organisms including numerous dicot and monocot angiosperm species and gymnosperms. In addition, *Agrobacterium* can transform fungi, including yeasts, ascomycetes, and basidiomycetes. Recently, *Agrobacterium* was reported to transfer DNA to human cells. The molecular and genetic basis for the host range of a given *Agrobacterium* strain remains unclear. Early work indicated that the Ti plasmid, rather than chromosomal genes, was the major genetic determinant of host range. Several virulence (*vir*) loci on the Ti plasmid, including *virC* and *virF*, were shown to determine the range of plant species that could be transformed to yield crown gall tumors. The *virH* (formerly called *pinF*) locus appeared to be involved in the ability of *Agrobacterium* to transform maize, as established by an assay in which symptoms of maize streak virus infection were determined following agroinoculation of maize plants. Other *vir* genes, including *virG*, contribute to the "hypervirulence" of particular strains.

**A. tumefaciens Causal Agent of Crown Gall Disease**

**Crown Gall disease caused by *Agrobacterium***

## Molecular Basis of *Agrobacterium*–Mediated Transformation T-DNA

The molecular basis of genetic transformation of plant cells by *Agrobacterium* is transfer from the bacterium and integration into the plant nuclear genome of a region of a large tumor-inducing (Ti) or rhizogenic (Ri) plasmid resident in *Agrobacterium*. Ti plasmids are on the order of 200 to 800 kbp in size. The transferred DNA (T-DNA) or Ri plasmid. T-regions on native Ti and Ri plasmids are approximately 10 to 30 kbp in size. Thus, T-regions generally represent less than 10% of the Ti plasmid. Some Ti plasmids contain one T-region, whereas others contain multiple T-regions. The processing of the T-DNA from the Ti plasmid and its subsequent export from the bacterium to the plant cell result in large part from the activity of virulence (*vir*) genes carried by the Ti plasmid. T-regions are defined by T-DNA border sequences. These borders are 25 bp in length and highly homologous in sequence. They flank the T-region in a directly repeated orientation. In general, the T-DNA borders delimit the T-DNA, because these sequences are the target of the VirD1/VirD2 border-specific endonuclease that processes the T-DNA from the Ti plasmid. There appears to be a polarity established among T-DNA borders: right borders initially appeared to be more important than left borders. We now know that this polarity may be caused by several factors. First, the border sequences not only serve as the target for the VirD1/VirD2 endonuclease but also serve as the covalent attachment site for VirD2 protein. Within the Ti or Ri plasmid (or T-DNA binary vectors), T-DNA borders are made up of double- stranded DNA. Cleavage of these double stranded border sequences requires VirD1 and VirD2 proteins, both in vivo and in vitro. In vitro, however, VirD2 protein alone can cleave a single- stranded T-DNA border sequence. Cleavage of the 25-bp T-DNA border results predominantly from the nicking of the T-DNA "lower strand," as conventionally presented, between nucleotides 3 and 4 of the border sequence. However, double-strand cleavage of the T-DNA border has also been noted. Nicking of the border is associated with the tight (probably covalent) linkage of the VirD2 protein, through tyrosine 29, to the 5' end of the resulting single stranded T-DNA molecule termed the T-strand. It is ssT-strand, and not a double- stranded T-DNA molecule, that is transferred to the plant cell. Thus, it is the VirD2 protein attached to the right border and not the border sequence per se, that establishes polarity and the importance of right borders relative to left borders. It should be noted, however, that because left-border nicking is also associated with VirD2 attachment to the remaining molecule (the "non-T- DNA" portion of the Ti plasmid or "backbone" region of the T- DNA binary vector), it may be possible to process T-strands from these regions of Ti and Ri plasmids and from T-DNA binary vectors. Second, the presence of T-DNA "overdrive" sequences near many T-DNA right borders, but not left borders, may also help establish the functional polarity of right and left borders. Overdrive sequences enhance the transmission of T-strands to plants, although the molecular mechanism of how this occurs remains unknown. Early reports suggested that the VirC1 protein binds to the overdrive sequence and may enhance T-DNA border cleavage by the VirD1/VirD2 endonuclease. *VirC1* and *virC2* functions are important for virulence; mutation of these genes results in loss of virulence.

## T-DNA Transfer from *Agrobacterium* to Plant Cells

As indicated above, many proteins encoded by *vir* genes play essential roles in the *Agrobacterium*-mediated transformation process. The role of Vir proteins is that they may serve as points of manipulation for the improvement of the transformation process. VirA and VirG proteins function as members of a two components sensory-signal transduction genetic regulatory system. VirA is a periplasmic antenna that senses the presence of particular plant phenolic compounds that are induced on wounding. In coordination with the monosaccharide transporter ChvE and in the presence of the appropriate phenolic and sugar molecules, VirA autophosphorylates and subsequently transphosphorylates the VirG protein. VirG in the nonphosphorylated form is inactive; however, on phosphorylation, the protein helps activate or increase the level of transcription of the *vir* genes, most probably by interaction with *vir*-box sequences that form a component of *vir* gene promoters. Constitutively active VirA and VirG proteins that do not require phenolic inducers for activity, or VirG proteins that interact more productively with *vir*-box sequences to activate *vir* gene expression, may be useful to increase transformation efficiency or host range. Together with the VirD4 protein, the 11 VirB proteins make up a secretion system necessary for transfer of the T-DNA and several other Vir proteins, including VirE2 and VirF. VirD4 may serve as a "linker" to promote the interaction of the processed T-DNA/VirD2 complex with the VirB-encoded secretion apparatus. Most VirB proteins either form the membrane channel or serve as ATPases to provide energy for channel assembly or export processes. Several proteins, including VirB2, VirB5, and possibly VirB7, make up the T-pilus. VirB2, which is processed and cyclized, is the major pilin protein. The function of the pilus in T-DNA transfer remains unclear; it may serve as the conduit for T-DNA and Vir protein transfer, or it may merely function as a "hook" to seize the recipient plant cell and bring the bacterium and plant into close proximity to affect molecular transfer. One aspect of pilus biology that may be important for transformation is its temperature lability. Although *vir* gene induction is maximal at approximately 25 to 27°C, the pilus of some, but not all, *Agrobacterium* strains is most stable at lower temperatures (approximately 18 to 20°C). Early experiments indicated a temperature effect on transformation. Thus, one may consider cocultivating *Agrobacterium* with plant cells at lower temperatures during the initial few days of the transformation process.

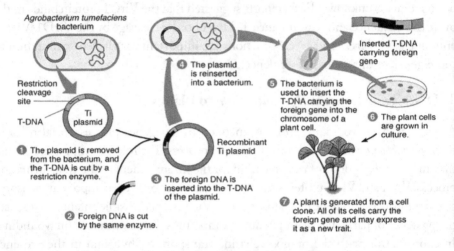

**Agrobacterium mediated genetic transformation**

The VirD2 and VirE2 proteins play essential and perhaps complementary roles in *Agrobacterium*-mediated transformation. These two proteins have been proposed to constitute, with the T- strand, a "T-complex" that is the transferred form of the T-DNA. Whether this complex assembles within the bacterium remains controversial. VirE2 could function in a plant cell: transgenic VirE2-expressing tobacco plants could "complement" infection by a *virE2* mutant *Agrobacterium* strain. Several laboratories have shown that VirE2 can transfer to the plant cell in the absence of a T-strand, and it is possible that VirE2 complexes with the T-strand either in the bacterial export channel or within the plant cell. A recent report suggests perhaps another role for VirE2 early in the export process: Dumas et al. showed that VirE2 could associate with artificial membranes in vitro and create a channel for the transport of DNA molecules. Thus, it is possible that one function of VirE2 is to form a pore in the plant cytoplasmic membrane to facilitate the passage of the T-strand. Because of its attachment to the 5' end of the T-strand, VirD2 may serve as a pilot protein to guide the T-strand to and through the export apparatus. Once in the plant cell, VirD2 may function in additional steps of the transformation process. VirD2 contains nuclear localization signal (NLS) sequences that may help direct it and the attached T-DNA to the plant nucleus. The NLS of VirD2 can direct fused reporter proteins and in vitro-assembled T-complexes to the nuclei of plant, animal, and yeast cells. Furthermore, VirD2 can associate with a number of *Arabidopsis* importin- proteins in an NLS-dependent manner, both in yeast and in vitro. Importin is a component of one of the protein nuclear transport pathways found in eukaryotes. Recent data, however, suggest that VirD2 may not be sufficient to direct T-strands to the nucleus. Ziemienowicz et al. showed that in permeabilized cells, VirD2 could affect the nuclear targeting of small linked oligonucleotides generated in vitro but could not direct the nuclear transport of larger linked molecules. To achieve nuclear targeting of these larger molecules, VirE2 additionally had to be associated with the T-strands. Finally, VirD2 may play a role in integration of the T-DNA into the plant genome. Various mutations in VirD2 can affect either the efficiency or the "precision" of T-DNA integration.

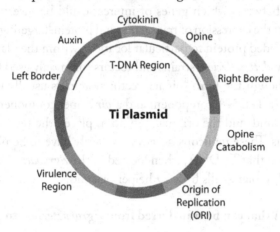

**Tumor inducing plasmid**

## Manipulation of *Agrobacterium* for Genetic Engineering Purposes

Ti plasmids are very large and T-DNA regions do not generally contain unique restriction endonuclease sites not found elsewhere on the Ti plasmid. Therefore, one cannot simply clone a gene of interest into the T-region. Because of the complexity of introducing foreign genes directly into the T-region of a Ti plasmid, several laboratories developed an alternative strategy to use *Agrobacterium* to deliver foreign genes to plants. The T-region and the *vir* genes could be separated into two different replicons. When these replicons were within the same *Agrobacterium* cell, products of the *vir* genes could act in *trans* on the T-region to effect T-DNA processing and transfer to a plant cell. Hoekema et al. called this a binary-vector system; the replicon harboring the T- region constituted the binary vector, whereas the replicon containing the *vir* genes became known as the *vir* helper. The *vir* helper plasmid generally contained a complete or partial deletion of the T-region, rendering strains containing this plasmid unable to incite tumors. A number of *Agrobacterium* strains containing non-oncogenic *vir* helper plasmids have been developed, including LBA4404, GV3101 MP90, AGL0, EHA101 and its derivative strain EHA105, and NT1 (pKPSF2). T-DNA binary vectors revolutionized the use of *Agrobacterium* to introduce genes into plants. Scientists without specialized training in microbial genetics could now easily manipulate *Agrobacterium* to create transgenic plants. These plasmids are small and easy to manipulate in both *E. coli* and *Agrobacterium* and generally contain multiple unique restriction endonuclease sites within the T-region into which genes of interest could be cloned. Many vectors were designed for specialized purposes, containing different plant selectable markers, promoters, and poly(A) addition signals between which genes of interest could be inserted, translational enhancers to boost the expression of transgenes, and protein-targeting signals to direct the transgene encoded protein to particular locations within the plant cell provide a summary of many *A. tumefaciens* strains and vectors commonly used for plant genetic engineering. Although the term "binary vector system" is usually used to describe two constituents (a T-DNA component and a *vir* helper component), each located on a separate plasmid, and the original definition placed the two modules only on different replicons. These replicons do not necessarily have to be plasmids. Several groups have shown that T- DNA, when located in the *Agrobacterium* chromosome, can be mobilized to plant cells by a *vir* helper plasmid.

## Amount of DNA that can be transferred from *Agrobacterium* to plants

The T-regions of natural Ti and Ri plasmids can be large enough to encode tens of genes. For example, the T-region of pTiC58 is approximately 23 kbp in size. In addition, some Ti and Ri plasmids contain multiple T-regions, each of which can be transferred to plants individually or in combination. For purposes of plant genetic engineering, scientists may wish to introduce into plants large T-DNAs with the capacity to encode multiple gene products in a biosynthetic pathway. Alternatively, the reintroduction of large regions of a plant genome into a mutant plant may be useful to identify, by genetic complementation, genes responsible for a particular phenotype. How large a T-region can be transferred to plants? Miranda et al. showed that by reversing the orientation of a T-DNA right border, they could mobilize an entire Ti plasmid, approximately 200 kbp, into plants. Although the event was rare, this study showed that very large DNA molecules could be introduced into plants using *Agrobacterium* mediated transformation. Hamilton et al. first demonstrated the directed transfer of large DNA molecules from *Agrobacterium* to plants by the development of a binary BAC (BIBAC) system. Others showed that a 150-kbp cloned insert of human DNA could be introduced into plant cells by using this system. However, the efficient transfer of such a large DNA segment required the overexpression of either *virG* or both *virG* and *virE*. VirE2 encodes a single-stranded DNA binding protein that protects the T-DNA from degradation in the plant cell. Because *virG* is a transcriptional activator of the *vir* operons, expression of additional copies of this regulatory *vir* gene was thought to enhance the expression of VirE2 and other Vir proteins involved in T-DNA transfer. Overexpression of *virE* formed part of the BIBAC system that was used to transform large (30- to 150-kbp) DNA fragments into tobacco and the more recalcitrant tomato and *Brassica*. However, the transfer of different-size T-DNAs from various *Agrobacterium* strains had different requirements for overexpression of *virG* and *virE*. Liu et al. developed a transformation competent artificial chromosome vector system based on a P1 origin of replication and used this system to generate libraries of large (40- to 120-kbp) *Arabidopsis* and wheat DNA molecules. This system did not require overexpression of *virG* or *virE* to affect the accurate transfer of large fragments to *Arabidopsis*.

## T-DNA Integration and Transgene Expression

Plant transformation does not always result in efficient transgene expression. The variable expression levels of transgenes, which frequently did not correlate with transgene copy number due to this lack of correspondence was initially attributed to position effects, i.e., the position within the genome into which the T-DNA integrated was credited with the ability of transgenes to express. T-DNA could integrate near to or far from transcriptional activating elements or enhancers, resulting in the activation of T-DNA-carried transgenes. T-DNA could also integrate into transcriptionally competent or transcriptionally silent regions of the plant genome. The high percentage (approximately 30%) of T-DNA integration events that resulted in activation of a promoterless reporter transgene positioned near a T-DNA border suggested that T-DNA may preferentially integrate into transcriptionally active regions of the genome. Only integration events that would link the promoterless transgene with an active promoter would result in reporter activity. However, a drawback to some of these experiments was that transgenic events may have been biased by the selection of antibiotic resistant plants expressing an antibiotic marker gene carried by the T-DNA. It is not clear whether T-DNA insertions into transcriptionally inert regions of the genome would have gone unnoticed because of lack of expression of the antibiotic resistance marker gene. An obvious way to circumvent the presumed problems of position effect is to integrate T-DNA into known transcriptionally active regions of the plant genome. An alternative system for gene targeting is the use of site-specific integration systems such as Cre-*lox*. However, single-copy transgenes introduced into a *lox* site in the same position of the plant genome also showed variable levels of expression in independent transformants. Transgene silencing in these instances may have resulted from transgene DNA methylation. Such methylation- associated silencing was reported earlier for naturally occurring T- DNA genes. Thus, transcriptional silencing may result from integration of transgenes into regions of the plant genome susceptible to DNA methylation and may be a natural consequence of the process of plant transformation. We now know not only that transgene silencing results from "transcriptional" mechanisms, usually associated with methylation of the transgene promoter, but also that transgene silencing is often "posttranscriptional"; i.e., the transgene is transcribed, but the resulting RNA is unstable. Such posttranscriptional gene silencing is frequently associated with multiple transgene copies within a cell. Although *Agrobacterium* mediated transformation usually results in a lower copy number of integrated transgenes, it is common to find tandem copies of a few T-DNAs integrated at a single locus. Transgene silencing can occur in plants harboring a single integrated T-DNA. However, integration of T-DNA repeats, especially head-to-head' inverted repeats around the T-DNA right border, frequently results in transgene silencing. Thus, a procedure or *Agrobacterium* strain that could be used to generate transgenic plants with a single integrated T-DNA would be a boon to the agricultural biotechnology industry and to plant molecular biology in general.

**Limitations as routine Ti plasmid vectors**

The production of phytohormones by transformed cells prevents them from being regenerated into mature plants.

A gene encoding opine synthesis is not useful to a transgenic plant and may lower the final plant yield

Ti plasmids are large (approximately 200 to 800 kb).

Ti plasmid does not replicate in *Escherichia coli*, therefore it cannot be cloned in *E. coli*.

Transfer of the T-DNA, which begins from the right border, does not always end at the left border. Rather, vector DNA sequences past the left border are often transferred.

## *Agrobacterium rhizogenes* meadiated biotransformation

**Hairy roots and Agrobacterium rhizogenes**

The causative agent of hairy root syndrome, is a common soil bacterium (Gram negative) capable of entering a plant through a wound and causing a proliferation of secondary roots. The underlying mechanism of hairy root formation is the transfer of several bacterial genes to the plant genome. The observed morphogenic effects in the plants after infection have been attributed to the transfer of part of a large plasmid known as the Ri (root-inducing) plasmid. The symptoms observed with *A. rhizogenes* are suggestive of auxin effects resulting from an increase in cellular auxin sensitivity rather than auxin production.

Ri plasmids are large (200 to greater than 800 kb) and contain one or two regions of T-DNA and a *vir* (virulence) region, all of which are necessary for tumorgenesis. The Ri-plasmids are grouped into two main classes according to the opines synthesized by hairy roots. First, agropine-type strains induce roots to synthesise agropine, mannopine and the related acids. Second, mannopine-type strains induce roots to produce mannopine and the corresponding acids. The agropine-type Ri-plasmids are very similar as a group and a quite distinct group from the mannopine-type plasmids. Perhaps the most studied Ri-plasmids are agropine-type strains, which are considered to be the most virulent and therefore more often used in the establishment of hairy root cultures.

The T-DNA of the agropine-type Ri-plasmid consists of two separate T-DNA regions designed the TL-DNA and TR-DNA. Each of the T-DNA fragments spans a 15 - 20 kb region, and they are separated from each other by at least 15 kb of non-integrated plasmid DNA. These two fragments can be transferred independently during the infection process. The genes encoding auxin synthesis (*tms*1 and *tms*2) and agropine synthesis (*ags*) have been localised on the TR-DNA of the agropine type Ri-plasmid. The mannopine type Ri-plasmids contain only one T-DNA that shares considerable DNA sequence homology with TL of the agropine-type plasmids.

Mutation analysis of the TL-DNA has led to identification of four genetic loci, designed locus *rol*A, *rol*B, *rol*C, and *rol*D, which affect hairy root induction. The complete nucleotide sequence of the TL-region revealed the presence of 18 open-reading frames (ORFs), 4 of which, ORFs 10, 11, 12 and 15, respectively, correspond to the *rol*A, *rol*B, *rol*C, and *rol*D loci.

One of the earliest stages in the interaction between *Agrobacterium* and a plant is the attachment of the bacterium to the surface of the plant cell. A plant cell becomes susceptible to *Agrobacterium* when it is wounded. The wounded cells release phenolic compounds, such as acetosyringone, that activate the *vir*-region of the bacterial plasmid. It has been shown that the *Agrobacterium* plasmid carries three genetic components that are required for plant cell transformation.

It has been shown that the *Agrobacterium* plasmid carries three genetic components that are required for plant cell transformation. The first component, the T-DNA that is integrated into the plant cells, is a mobile DNA element. The second one is the virulence area (*vir*), which contains several *vir* genes. These genes do not enter the plant cell but, together with the chromosomal DNA (two loci), cause the transfer of T-DNA. The third component, the so-called border sequences (25 bp), resides in the *Agrobacterium* chromosome. The mobility of T-DNA is largely determined by these sequences, and they are the only *cis* elements necessary for direct T-DNA processing.

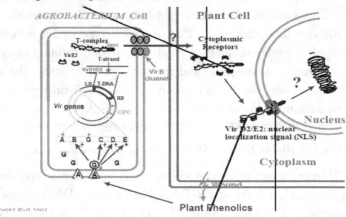

***Hairy roots*** are fast growing and laterally highly branched, and are able to grow in hormone-free medium. Moreover, these organs are not susceptible to geotropism anymore. They are genetically stable and produce high contents of secondary metabolites characteristic to the host plant. The secondary metabolite production of hairy roots is stable compared to other types of plant cell culture. The alkaloid production of hairy roots cultures has been reported to remain stable for years. The secondary metabolite production of hairy roots is highly linked to cell differentiation. Alkaloid production decreased clearly when roots were induced to form callus, and reappeared when the roots were allowed to redifferentiate. An interesting characteristic of some hairy roots is their ability to occasionally excrete the secondary metabolites into the growth medium. However, the extent of secondary product release in hairy root cultures varies among plant species.

The average growth rate of hairy roots varies from 0.1 to 2.0 g dry weight/liter/day. This growth rate exceeds that of virtually all-conventional roots and is comparable with that of suspension cultures. However, the greatest advantage of hairy roots compared to conventional roots is their ability to form several new growing points and, consequently, lateral branches. The growth rate of hairy roots may vary greatly between species, but differences are also observed between different root clones of the same species. The pattern of growth and secondary metabolite production of hairy root cultures can also vary. Secondary production of the hairy roots of *Nicotiana rustica* L. was strictly related to the growth, whereas hairy roots of *Beta vulgaris* L. exhibited non-growth-related product accumulation. In the case of the hairy roots of *Scopolia japonica* Jacq. and *H. muticus*, the secondary products only started to accumulate after growth had ceased. Secondary metabolite synthesis dissociated from growth would be desirable for commercial production, as it would allow the use of continuous systems.

## 1. Induction of hairy roots via *A. rhizogenes* Infection:

Surface-sterilized cotyledons are wounded and infected with *A. rhizogenes* strains. The inoculated cotyledons are co-cultivated with *A. rhizogenes* strains for 2 days at 25C with a 16 hours photoperiod. The experiment is designed to be completely randomized with four replicates. Forty explants are used for each population. After co-cultivation, explants are transferred to hormone-free growth mediums (High salt media such as MS favors hairy root formation in some plants. Low salt media such as B5 favor excessive bacterial multiplication in the medium and therefore the explant needs to be transferred several times to fresh antibiotic containing medium before incubation.), semi-solid MS (Murashige and Skoog) medium solidified with 0.8% agar, and contained 3% sucrose, plus 0.4g/l augmentin to kill the bacteria (pH:

5.7) at a density of 10 explants per plate (9 cm petri dish), and cultured at 25C, with a 16 hr photoperiod.

Frequency of hairy root formation for each treatment is scored 30 days after co-cultivation. Individual roots will emerge from the wound sites, they are excised and sub-cultured onto the same medium. Forty days after co-cultivation, hairy roots are weighed out and transferred to 50 ml of MS liquid medium (pH= 5.7) containing 3% sucrose, and shaken in an orbital shaker at 120 rev/min at 25°C in the dark. The roots are then sub-cultured onto the same medium every 4 weeks. After 4 months in liquid culture, hairy roots from each explant are weighed out and the mean weight for each treatment iss calculated.

## Culture conditions

1. The susceptibility of plant species to *Agrobacterium* strains varies greatly. Significant differences were observed between the transformation ability of different strains of *Agrobacterium*.

2. The age and differentiation status of plant tissue can affect the chances of successful transformation.

3. The level of tissue differentiation determines the ability to give rise to transformed roots after *A. rhizogenes* inoculation. In this case, successful infection of some species can be achieved by the addition of acetosyringone.

## Detection of Biotransformation

### 1) Determination of Anthraquinone Contents; how much opine produced in hairy roots.

Hairy roots were dried in the dark at 60°C for 2 days. Dried roots were powdered by mortar and pestle, and 50 mg of this fine powder was then soaked in 50 ml of distilled water for 16 h. This suspension was heated in water bath at 70°C for 1 h. After the suspension was cooled, 50 ml of 50% methanol (MeOH) was added and then filtrated. The clear solution was measured by spectrophotometer (Shimadzu UV-160A) at a wave length of 450 nm and compared with a standard solution containing 1mg/100ml alizarin and 1 mg/100ml purpurin with the absorption-maximum 450 nm. However, there is a disadvantage that opines production can be unstable in hairy roots and may disappear after a few passages.

## 2) T-DNA detection by southern blot hybridization

Genomic DNA from transformed and non-transformed soil-grown plants were extracted using the CTAB extraction method. Approximately 10 mg of DNA from each sample were digested with HindIII, BamHI, EcoRI, respectively, non-transformed plant was digested with EcoRI, then separated by electrophoresis 0.8% (w/v) agarose gel, transferred from the agarose gel to Hybond+ nylon membrane and cross-linked to the membrane by UV light for 3 min. Probe were labeled with a 32P labeled probe specific to the coding sequence of the introduced *rolB*, or rolA, or rolC gene for Southern hybridization. Filters were pre-hybridized in 5 $_{i \times}$SSC, 5 $_{i \times}$Denhardt's solution, 0.5% SDS, 20 mg/ml denatured salmon sperm DNA at 65oC and subsequently hybridized overnight with labeled probe. After stringent washing (0.1 $_{i \times}$SSC, 0.1% SDS, 65oC) filters were autoradiographed at -70oC for 3 days with an intensifying screen.

## 3) Bacterial gene detection by PCR

The polymerase chain reaction was used to confirm the presence of rolB, or rolA, or rolC gene in roots by their primers. The PCR reactions were carried out in a total volume of 30 ml: 1 ml samples of the transformed plant genomic DNA, 20 pmol of each primer, 200 m *M* each dNTP, 0.5 units Taq DNA polymerase and 3 ml 10 $_{i \times}$PCR buffer. Cycling conditions were: denaturation at 94oC for 1 min, annealing at 55 oC for 1 min and extension at 72 oC for 3 min. Samples were subjected to 30 cycles. Amplification products were analyzed by electrophoresis on 0.8% agarose gels and detected straining with ethidium bromide.

## Advantages of Hairy root cultures

Hairy roots also offer a valuable source of root derived phytochemicals that are useful as pharmaceuticals, cosmetics, and food additives. These roots can also synthesize more than a single metabolite and therefore prove economical for commercial production purposes. Transformed roots of many plant species have been widely studied for the *in vitro* production of secondary metabolites. Transformed root lines can be a promising source for the constant and standardized production of secondary metabolites. Hairy root cultures produce secondary metabolites over successive generations without losing genetic or biosynthetic stability. This property can be utilized by genetic manipulations to increase biosynthetic capacity.

# Chapter - 14

## Bioreactors Used in Plant Tissue Culture

**Introduction to plant cells:** The use of plant cell and tissue cultures for production of biologically active substances is called plant cell-based bioprocessing; the active substances that this technique allows was low molecular secondary metabolites and recombinant proteins. Plant cell-based bioprocessing has some significant advantages over the traditionally grown of the whole wild plant or transgenic plant. The most important advantage is the sterile production of metabolites under defined controlled conditions, this means that climatic changes and soil conditions are not able to influence the product yield and quality.

Plant cells have plenty of advantages over mammalian cells, insect cells, and bacteria; these cells are capable of performing complex posttranscriptional processing, which is a precondition for heterologous protein expression. When they are compared with mammalian cells (which dominate the commercial protein manufacture) plant cell cultures have lower cost and are safer due to the lower risk of contamination by viruses, pathogens and toxins. Also plant cells grow at room temperature (25-27°C), need a lower aeration (0.1-0.3 vvm) than microbial culture and they have a lower shear stress than mammalian cells due the cell wall. Compelling disadvantages which scientists have to deal, are the difficult scale-up, the low cell density, the intracellular product accumulation (mostly in vacuoles), but the most significant disadvantage is the slow growth rate with a $t_d$ ( division time ) of 2 to 7 days.

**Plants cells characteristics:** Eukaryotic systems, like plant cells, have the ability to produce secondary metabolites and glycoproteins, which the pharmaceutical industry uses for creates recombinant proteins and secondary metabolites which are used in both industries pharmaceutical and cosmetic. Sigmoid curve ideally represent the cell growth, with lag, exponential, delay, stationary growth and lethal phases. In addition, plant cells have totipotency (potential to form all cell types and a whole plant), as is said before, plant cells have a doubling times of days to weeks due to their large size ( 100 μm). The robustness of plant cells is moderate and can be attributed to the cell wall, this robustness have some variations in order to age of the culture, species and culture type, they have lower shear stress, and can be cultivated in reasonable agitation and aeration conditions.

**Culture conditions:** Mostly of the plant culture cells have similar culture conditions as temperature, between 25 and 27°C, a medium pH with an optimal parameter of 5.0 and 6.0. Also culture cells need some aeration; this aeration is extremely lower than microbial systems and comparable to mammalian cultures. Some cultures have periodic dark/light cycle of 8 h and 16 h or continuous introduction of light (0,6 -10 klux or 80,7-1345 μmole $m^{-2}$ $s^{-1}$ ), mammalian cells have to be cultivated in dark as the consequence of the light-sensitive compounds of their media culture. In order to eliminate long lag phases (120 hours) plant cell cultures can be initiated with high cell concentrations (10% of culture volume).

**Media:** Nutrient supply is a critical element when culturing plant cells. The main substance is doubled distilled and deionized water which represents 95% of the media. All culture media have a basal medium with carbon source like fructose, sucrose, glucose and sorbitol, also with organic supplement and inorganic supplements; in organic supplements as amino acids, vitamins and cofactors as tocopherol are used. The inorganic supplements are macro and microelements, the microelements are in μM concentrations and macroelements in mM. In addition nutrient supply needs some phytohormones as growth regulators like auxins, cytokinins and gibberellins. Growth regulators affect the growth process, cytokinins such as adenine which promotes cell division and auxins, like indole-3-acetic acid (IAA), or 2,4-dichlorophenoxyacetic acid (2,4-D) which is used as a dedifferentiating hormone for rapid callus induction. With low auxin concentration and a high cytokinin concentration the cell growth are stimulated and switched concentrations cell division are promoted. Recombinant protein production is influenced by supplements which stabilizes the proteins like bovine serum albumin, PVP (polyvinylpyrrolidone), gelatin and sodium chloride. These components are used in medium MS developed by Murashige and Skoog for tobacco cultures, among others. Finally if a solid support matrix is needed, the addition of agar, agarose, and gellan gums can reach this purpose.

**Cultures:** Industry and research have different culture types such as callus, plants cell suspension, hairy root cultures, embryogenic and shoot cultures

**Callus cultures:** Callus cultures are fundamental in vitro culture of plants cells, not only for acquire a high density of biomass but also because they are necessary for the establishment of cell suspensions.Callus is described by an amorphous mass of unorganized parenchyma cells, the formation of callus normally is achieved by placing the explant on an appropriate solid growth medium with the necessary components such as phythotmones in a Petri dish incubated at 25°C in darkness or low light. Callus cultures have a developmental stage of cells described as dedifferentiated cells (DDC, restored totipotency). A DDC callus has friability, a soft cell mass with fast exponential growth. If a hard callus is achieved, Evan's method can reestablish a friable callus with addition of phytohormons like auxins, pectinase or shorter subculturing intervals. Callus has particular maintenance and culture conditions, with a doubling time around 7 days; the subcultivation is necessary every 3 to 4 weeks (depending on species).

**Hairy roots:** Hairy roots (transformed rots) are generated by the transformation of plants or explants with agropine and mannopine-type strains of *Agrobacterium rhisogenes* (a gram negative bacterium). Neoplastic roots are the developmental stage of cells and roots covered with tiny hairs are the appearance, a medium liquid or solid can achieve high densities of this culture. With a doubling time between one to eight days a subcultivation every two to four weeks is necessary. Hairy roots are genetically and biochemically very stable with a high integrity, but this type of culture has a shear-sensitive tissue. Consequently, it is difficult to have cultures in vitro cultivation (is required a special bioreactor design) and are also complicated the scale-up procedures.

**Embryogenic and shoot cultures:** Embryogenic and shoot cultures are used for micro propagation and plant breeding, and like hairy rots, they belong to the group of differentiated organ cultures. These kind of cultures are stablished via meristem, seed germination or embryonic culture.

**Plants cell suspension:** Plant cells grow in plant cell suspension and also, the growth of a callus is necessary for this type of culture. The callus produced is dispersed by inoculating the fragments of callus in a liquid medium, and if it's necessary to achieve friability in the callus, the Evan's method is required. After callus isolation and subcultivation the culture is usually filtered with a sieve of 0,3mm to 0, 5 mm in order to remove large aggregates. Homogenization aims to produce a well-mixed suspension. Established plant cell suspension is generally maintained in shake flasks at 25°C and between 110 to 120 rpm. With a doubling time of 0.6 and 5 days plant cell suspensions grows faster than callus culture. Plant cell suspensions are highly heterogeneous as variability in terms of morphology (cell size, cell shape, and cell aggregation), rheological characteristics, growth and metabolic pattern. Rheological characteristics are important in the process of scale-up; plant cell suspensions are non-newtonian fluids (with plastics and pseudoplastics characteristics), so the density of this cultures raises proportional to cell density. As a result plant suspension cells very rarely grow as single cells, they form aggregates based on cell adhesion and results from the secretion of wall extracellular polysaccharides, which prevent cell separation.

**Transformed plant cell cultures:** Cell lines of plants express the specific transferred gene. These cell lines are usually generated by Agrobacterium-mediated stable transformation. Now non-transformed plant cell cultures are used only if the natural biosynthesis pathway of the target compound has to be examined or if the target compound is naturally synthesized or expressed in a reasonably high amount. Transformed cell cultures, exist in order to optimize the production of a target compound with a stable transformation (all the generations of this plants have the same expression) or with a transient transformation (a few generations have the transformed form).

**Examples of different culture modes:** Not only bioreactor type is important to achieve a high density of product, also it is necessary to determine the culture mode; batch, fed-batch and continuous cultures such as chemostat perfusion and dialysism are some examples of different culture modes.

**Batch cultures:** Batch processes constitutes the classical bioreactor mode, infrequently applied in cell culture processes but used in microbial bioreactors such as fermenters. A closed system with a constant culture volume is harvested at maximum cell density or product titre. A few advantages of this mode is the simplicity and flexibility; these characteristics lead to an easy scale-up, without contamination, are cheaper and have shorter or less steps. But low space-time yield and problems in process characterized by inhibition of substrate and product (less production) are some of the most important disadvantages of this mode and make him useless.

**Fed-batch cultures:** This kind of batch is often used in industrial applications; with this system, industries can reach higher living cells densities and product titres compared to batch cultivations. Exist three different versions: One with a culture volume increased continuously (feeding), other with a culture volume kept constant so using and exchange and feeding, and finally a repeated fed batch this one after fed batch-phase, cell suspension is partially harvested and fresh medium is fed to the system. The main advantage of this model is the higher cell concentration that can reach and the no limitation of high production, but is complex, expensive and the system has to be developed.

**Continuous culture:** Is an opened system with a constant culture volume and the process runs in a batch mode up to desired cell density which differs between different species. The subsequent introduction of continuous mode is based on a continuous medium feed and withdrawal of culture broth. Two different processes can be used in continuous batch, without cell retention and with cell retention, with cell retention is called perfusion. Different advantages such as high space-time yields (months) and smaller bioreactors, lower investment costs for processes with cell retention; never cell inhibition by product (extracellular) or substrate happens and present high sterile environment. In the other hand needs a higher levels of instrumentation, higher level of consumption of culture medium and costs of down-streaming, also exist a probable wash out of cells (elutriation) or nutrient limitations when dilatation rate is too high or too low and finally more complicated validation and registration procedures. Also it is laborious, and has a high duration.

**Bioreactors:** A bioreactor is defined as a closed system (vessel/bag or apparatus) in which a biochemical reaction involving biocatalyst takes place. In the process the biocatalysis is converted into an expressed protein which is biomass or expressed proteins, it should be pointed out that the term "fermenter" is used only by bioreactors which involves fast growing microorganisms, but in American English this term is used in both bioreactors. The primary role of a bioreactor is to provide containment with sustainable conditions for cell growth and/or product formation. In general a cell culture bioreactor has to meet the following demands: Guaranteed cell-to-cell contact and a surface for cell detachment in case of anchorage-dependent growing cells.

- » Homogeneous and low-shear mixing and aerations.
- » Sufficient turbulence of effectual heat transfer.
- » Adequate dispersion of air and gas.
- » Avoidance of substrate segregation.
- » Measurability of process variables and key parameters.
- » Scale-up capability.
- » Long-term stability and sterility.
- » Easy of handling.
- » Reasonable maintenance.
- » Avoids all possible contamination of the culture.

Before describing all the bioreactors it's imperative to know a few characteristics about how to determinate grow of the plant cells and critical culture points in this kind of devices.

**In-Process control (IPC), determination of plant cell growth:** Process monitoring and control are generally facilitated by standards bioreactor facility which uses pressure, temperature, pH gas flow rate, pO2 (dissolved oxygen), pCO2 (dissolved carbon dioxide), and conductivity sensors. Not only in bioreactors IPC is required, also in maintenance of plant cell cultures due to the high quality cell culture material which is constantly available, this maintenance is carried out with high viable cells (<90%) at regular time intervals. A detailed knowledge of the plant cell culture is required such as inoculum density or if cells are genetically modified or not. To characterize and design bioprocesses based on plant cells with secondary metabolite production or protein accumulation/ secretion and nutrient utilization, cell growth must be determined. Plant cell growth must be determined in both processes such as experimental amounts or in scale-up productions, normally is measured as fresh weight or as dry weight. Fresh weight values (expressed in g) are obtained by weighing freshly harvested cells. Dry weight avoids errors caused by endogenous water content (accumulated in the vacuole) and is more useful tool for biomass quantification than fresh weight. Dry weight measurement uses a known weight of fresh plant cells dried in an oven at a temperature between 50-60°C during 24 to 48 hours. Also fresh cells can be lyophilized.

In addition morphology and viability can be determined by microscope using Evan's blue, a dye which is used for staining suspended cell counts. Viability is defined as the ratio of viable cells to total cells and has a value between 0 and 100, described as 0% to 100%, blue cells are dead cells and the yellow ones are alive. [Figure 3]. Viability is a subjective determination, so two or three people are necessary. For cell suspensions like BY-2, another important parameter is determined, this parameter is called Packed Cell Volume or PCV, acquired by gentle centrifugation at low speed. PCV defines the ratio between the volumes occupied by biomass to the volume of the whole sample (aliquots of 10 mL). Biomass and packed cell volume have a correlation; also biomass can be measured by Neubauer type hemocytometer doing a manual counting.

Indirect monitoring is also a reliable measurement technique, consisting on measurements such as conductivity with a conductivity meter in liquid culture medium which allows the indirect monitoring of biomass growth. There exists an inverse correlation between electrical conductivity (expressed as mS cm$^{-1}$) and biomass. This can be explained by the uptake of nutrients and salts by the cells.

Graphic 2 and 3 expresses this correlation. The pH measurement is routinely made during the cultivation of liquid cultures. A gradual drop in pH to a value around 4, reflects the initial ammonium uptake and acidification caused by cell lysis within 20-48 hours as is possible to see in graphic 3. The pH returns to a stable value of about 5 related to an uptake of nitrates after a few days of cultivation, finally at the end of the culture, medium reaches a pH above 6. Metabolites are measured also, such as sucrose, glucose, fructose, ammonium, nitrate and phosphate. In graphic 4 it's possible to observe that sucrose is the first metabolite being consumed by the cells and after that it is observed the raises of glucose and fructose due to the fact that sucrose is metabolized in this two components. Ammonium and phosphate are important metabolites due to the possibility of inhibition of the cell culture due to their amount.

**Illumination:** External illumination is an important parameter, with a range of 0-10klux it's possible to obtain a periodic cycle between 8h to 16 h or continuous illumination, by fluorescent lamps, light-emitting diodes like LEDs or Solid State Lamps (SSL).Furthermore, it's possible to use an internal illumination with encapsulated fluorescent lamps, and fiber optic cables. This illumination is needed for certain kinds of cell cultures for grow or for elicitor but causes general problems such as heat development, and problems in the distribution of light intensity. In cosmetic industries, illumination is not IPC, this belongs to a dark cultures due to the low interest of the expression of chlorophyll which can dye with color green the product.

**Types of bioreactors for plant cell suspension culture:** For plant cell suspensions bioreactors can be divided into three main types according to their continuous phase: liquid phase bioreactors, gas-phase bioreactors and hybrid bioreactors.

**Liquid-phase bioreactors:** Plants cells are immersed continuously and oxygen is usually supplied by bubbling air through the culture medium. **Mechanically driven bioreactor, pneumatically driven bioreactor** and **hydraulically driven bioreactor** belong to this category. Due to the low solubility of gases, the gas-exchange limitation and insufficient nutrient transfer, growth inhibition may occur in this kind of cultures

**Gas-phase bioreactors:** In these bioreactors, oxygen transfer limitation can be reduced or even eliminated, with bioreactors such as mist reactor or spray reactor These reactors are specially manufactured for organ cultures like hairy roots, in addition a less hydrodynamic stress is acquired with this type of reactor.

**Hybrid bioreactors:** With a combination of submerged and emerged bioreactors such as Wilson Bioreactor for hairy roots, the hybrid bioreactor switches from liquid-phase to gas-phase operation after the inoculation, distribution, attachment to immobilization points, and short growth phase of the cells. An excellent plant cell growth with biomass productivity > 1g dry weight/L/day, requires and optimized and well characterized bioreactor configuration. Plant cells with high cell density and/ or aggregate formation. Presuppose trouble-free inoculation, transfer and harvest, and consequently need specially designed bioreactor elements. Mechanically or pneumatically driven aerated submerged bioreactors, are more often used in cell suspension cultures for large-scale rates. Derived from microbial fermenters stirred reactors, bubble column reactors and airlift reactors were initially used with only minor modifications to grow plant suspension cells. In most plant cells cultivations, the air is directly introduced via a sparger (ring, pipe, plate, frit) positioned in the lower part of the bioreactor. Such direct aeration guarantees the highest possible aeration efficiency.

**Re- and multi usable bioreactors for plant cell suspension culture:**

**Stirred:** Independent of scale, production organism type, and product, stirred cell culture in which power input for mass and heat transfer is controlled mechanically, dominate. Since the end of 1950s: stirred bubble column and airlift bioreactor predominate over the other types of bioreactors, stirred dominate and are preferable for biomass productivity exceeding 30g dw $L^{-1}$. This kind of bioreactors permits a maximum culture volume of 70m$^3$ with a H:D= 2:2 to 3:1 , often they are equipped with a bubble aeration, usually a sparger rings, with a 0.1-0.5 vvm and a $k_L{}^*a \geq 10$ hr$^{-1}$. Large slow moving axial flow impellers with tip speeds up to 1.5 ms$^{-1}$ such as marine impellers, special pitched blade impellers, spiral stirrers, helical ribbon impellers and anchor impellers. Also Rugston impellers suitable for limited applications or impellers with an improved design like concave blades.

**Bubble columns and airlift reactors:** Pneumatically driven systems do not specifically require the use of immobilized cells as they were developed for free suspension cells. In bubble columns and airlift bioreactors, mass and heat transfers is mostly achieved by direct sparging of a tall column with air or gas that is injected by static gas distributors (diffuser stones, nozzles, perforated planes diffuser rings) or dynamic gas distributors such as slot nozzles, Venturi tubes, injectors or ejectors. While ascending gas bubbles cause random mixing in bubble columns, fluid circulation in airlift bioreactors is obtained by a closed liquid circulation loop, which permits highly efficient mass transfer and improved flow and mixing. In airlift bioreactors, this circulation loop results from the mechanical separation of a channel for gas / liquid up flow and down flow. Both airlift bioreactors with an external loop and with an internal loop are available.

Due to their relatively simple mechanical configuration bubble columns and airlift bioreactors are characterized by low cost in comparison with stirred bioreactors and also, the lower energy requirement, minimizes problems of scale-up. However this bioreactors have a potential disadvantages such as variations in biomass concentration, viscosity, surface tension, ionic concentration, inadequate mixing, foaming, flotation, and bubble coalescence.

With a maximum culture volume of 20 $m^3$, and a H:D equal to 6:1 to 14:1, bubble columns take advantage in cultures which biomass productivities exceeds > 10 g dw $L^1$. Bubble columns and airlift reactors are less often used than stirred bioreactors due to the limitations at high biomass productivities of heterogeneous distribution and oxygen transfer. In addition if high aeration rates are achieved with a sparger rings, extensive foaming flotation and wall growth phenomena may occurs. Moreover, bursting gas bubbles in bubble columns or in airlift bioreactors can raise the shear or hydrodynamic stress damage. Scientifics designed potential upscaling issues such as ceramic or sintered steel porus spargers, with an external aeration with a bubble-free aeration using an oxygen enrichment with only $O_2$ at lower aeration rates, and changes in geometry, an example of this change is in the case of balloon-type bubble bioreactors.

**Hollow fiber:** Hollow fiber bioreactors belongs to hydraulically driven devices, it reaches a high density with a tissue-like architecture. However hollow fiber reactors gain perfusion mod, this belongs to the hydraulically driven systems where energy input is produced by special double-phase pumps. Cells grow in the extra-capillary space of thousands of fibers that have been potted into a cylindrical cultivation module. An oxygen enriched medium flows continuously thought the fibers; this kind of bioreactor is often used by cell growth which produces secreted proteins. This bioreactor can be used by adherent or suspension cells, both are optimal for this device, however, besides probable mass transfer limitations there is a lack of data in process monitoring in the immediate cell environment. Also exist a risk of product contamination by cell fragments and cell lysis products as well as of destruction of sensitive proteins by the high residence time of cells. But the main disadvantage of this bioreactor is their small culture volume, which ranges only from 2,5 to 1000L. In conclusion, hollow fiber bioreactor is used for fast, flexible, small-scale production of antibodies for diagnostic and research.

**Bed bioreactor:** Also a hydraulically driven bioreactor, more directly linked to the use of cells for cultivation in an immobilized form. Cells are immobilized on microcarriers (small particles, usually spheres from 100 to 300µm in size). Bed bioreactors can also be described in to main types, packed or fixed bed bioreactors and fluidized bed bioreactors. The fixed bed bioreactors have a high density packet carrier material, which forms a fixed bed. This device is composed of cylindrical bioreactors chamber filled with carriers of porous glass or macroporous materials, a gas exchanger, a medium storage tank, and a pump that circulates the culture medium between the bioreactor and the medium storage tank. Severus disadvantages such as poor gas transfer and detachment, and cell washout limit the applications of this bioreactor, although have a low surface shear rate, absence of particle-particle abrasion and an increased space-time yield.

The packed bed bioreactor operates in up flow mode, the bed expands at high liquid flow rates and follows the motion of the microcarriers to which cells have been attached. To optimize the mass and heat transfer, fluidized bed bioreactors operation aims to provide a fluidized bed to ensure movement of all particles and avoid sedimentation or flotation.

**Orbitally-shaken bioreactors:** Orbitally shaken single-use reactors are promising reactors, this affirmation it's possible due his characteristics, such as a simple and cost-efficient, because no complex built-in elements such as baffles or rotating stirrers are required. Also the liquid distribution induced by orbital shaking is well-defined and accurately predictable. And finally, the scale-up from small-scale systems, where shaken bioreactors are commonly applied, is simple and has been successfully proven up to the cubic meter scale. However they have some disadvantages, orbitally shaken single-use reactors are only suitable for certain applications such as cultivating animal or plant cells with low oxygen demand. Also orbitally shaken bioreactors can be performed in single-use conformation

**Single-use and disposable bioreactors for plant cells and tissue cultures:** Cultivation bags or rigid cultivation containers for single-use are used in these bioreactors. They are manufactured from different kinds of polymeric materials and they must be sterilized by gamma radiation, customized and validated. Different types of bioreactors are suitable for plant cells cultures and root cultures such as mechanically driven and pneumatically driven. Mechanically driven like orbitally shaken, wave-mixed bioreactors, stirred bioreactor, bioreactor with vertically oscillating perforated disk are different systems for these kind of bioreactors. Instead of the large number of mechanically bioreactors pneumatically bioreactors only exist in driven bubble columns.

**Life-reactor (Osmotek):** Is the first disposable bioreactor for plants cells described in literature; with a capacity of 1.5 to 5 liters; it is a pneumatically driven bubble column. Used successfully for production of embriocrops such as potatoes, bananas, ferns and gladiolas.

**Plastic-lined bioreactor:** A bubble column integrated in a plastic bag was the basic idea for the Life bioreactor, with a capacity between 28,5 to 100 liters, which is a pneumatically driven bubble column.

**Wave-mixed bioreactors:** Mechanically driven bioreactor with a capacity between 1L to 500L, is a system for different vendors, and they have differences in rocking motion ( 1D,2D or 3D), bag geometry and size instrumentation. These reactors have a plastic disposable cultivation chamber; the gas-permeable and surface-aerated bag is fixed by a clamp arrangement and moved on the rocker unit. Mass and energy transfer is manually adjusted via the rocking angle, rocking rate, and filling level. They present multiple advantages such as bubble- free surface aeration, well-investigated, uniformity of energy dissipation and negligible foaming, this means a low shear stress and an increase of oxygen transfer. This bioreactor is consequently a suitable bioreactor for cell growth, in contrast, they have some problems with rheological issues of plant cultures.

» Commercial examples: Wave Bioreactor system 500/1000, AppliFlex, Cell-tainer single-use bioreactor, BIOSTAT® CultiBag RM, Tsunami Bioreactor. The BIOSTAT® RM

» A fully GMP compliant, single-use, wave-mixed bioreactor and single use Flexsafe® bags are proven for a broad range of different cell lines incl. CHO, NS0, SF9, E.coli and mesenchymal stem cells. It takes the advantage of an excellent cell growth and robustness, high type of supply and consistent quality and easy to use rocker with advanced control capabilities.

» AppliFlex® - Designed by Applikon® biotechnology, the Appliflex® bioreactor consists of a 10-liter, 20-liter and 50-liter bioreactor bag designed for single-use. It´s especially manufactured for mammal cells like CHO and for cell lines of insects like sf-9 and sf-12.

Other bioreactor developed and tested by Nestle are also a suitable instrumentation for plant cell and tissue cultures, such as Wave and Undertow Bioreactor (WUB) a mechanically driven layer with a capacity between 10 to 100 L with only a mobile part which gets up and down in order to make waves. Slug Bubble Bioreactor (SBB) a pneumatically driven with a capacity of 10 to 70 L, with a slog flow regime but this bioreactor has a handicap: the bubble flow is not homogeneous. Finally also developed by Nestle exist the Simple immersion bioreactor Box-in-bag Temporary Immersion Bioreactor.

**Protalix bioreactor:** It is a pneumatically driven bubble column bioreactor with a high capacity over than 400 L, restricted by GMP-production.

**Large-Scale disposable shaking bioreactors:** These bioreactors are cylindrical vessels of 20 L and 50 L mounted on a standard RC-6 shaking machine). Advantages of this single-use bioreactor over others disposable devices: it's easy to use, have a well-defined gas/liquid mass transfer area , a low levels of hydromechanical stresses due to homogeneous distribution of the power consumption and most important, the reduction of initial costs (no need to purchase special bags and rocking machines)

**Orbitally-shaken bioreactors:** Defined at Re- and multi usable bioreactors

**Re- and multi usable bioreactors for root culture:** Liquid-phase, gas-phase and hybrid bioreactor systems are all suitable for growing root cultures: bubble columns, airlift bioreactors and mist bioreactors with sterile baskets installed. They have a maximum culture volume of 10 m³ (balloon type bubble bioreactors). On the other hand, cultivation of roots has some difficulties such as varying root thickness, root length, number of root hairs and different root branching sequence that produce a non-homogenous growth and production. At the same time these cultures have tendency to form clumps inherently composed of primary roots and their bridged lateral roots. Root has been more damaged by shear stress so mostly of this cultures need an isolation of the roots from the impeller. This kind of cultures must be immobilized, which promotes root growth, this is acquired with horizontal or vertical meshes and cages or with polyurethane foam.

**Mist reactor:** Mist bioreactor is a gas-phase instrument with the highest potential for the cultivation of hairy roots in gas-phase bioreactors. With a droplet generator the roots have an optimal growth if critical droplet size is between 1μm and 35 μm since in this case liquid nutrients are homogeneously distributed and the gas transfer into the rots is free of limitations such as oxygen stress. The cultivation container has an immobilization support with horizontal and vertical meshes.

**Low Cost Mist Bioreactor ( LCMBs):** Is the largest gas-phase bioreactor with a capacity of 60L. Low cost mist bioreactors were designed to grow *Artemisia annua* transformed roots and *Dianthus caryophyllus* shoots. The reactors use similar mist generators but the culture chambers were modified to meet the requirements of each application.

**Wilson bioreactor:** Is the largest hybrid bioreactor system with a capacity of 500L. The spray reactor reminds a gas-phase bioreactor with a cultivation container with horizontal meshes. This bioreactor has some problems such as very laborious handling and often associated with contamination issues, now is no longer in use.

**Balloon-type:** Is a spherical-glass bioreactor, which is used for root culture due their advantages of easy moisture and simplicity, although this assets, are no-longer in use due to easy contamination and difficulty of take out the roots. Useful for extracellular products.

**Single use vs re- and multiusable bioreactors. Advantages and disadvantages:** In R+D biofactories have different trends than commercial industries, they need a scale-down bioreactors in a small scale laboratory, multifermenters systems equipped with stirred bioreactors or bubble columns which acquires a high amount of data in less time, such as bench top bioreactors with industrial control mode, low cost and disposable bioreactors. Disposable or single-use bioreactors have different advantages over typical bioreactors, such as short time for production, lower costs, and high safety. For these reasons these bioreactors are used in the emergent industry. Furthermore a high flexibility and simplicity also sterilization is not needed, neither cleaning times. Although, single-use bioreactors are well established, not always are optimal, they have some weakness such as limited scalability, limited standardization, an increased storage requirements and increase of waste generation (plastic bags have to be throw away) also a repetitive costs and security risks at supply the bioreactors with the mixture of culture and substrate. Now, some possible threats to increase production and acquire low costs and defeat the supplier dependence are possible with an union with the supplier; or with improved films as the reduction of Irgafos 168 in the new improved films.

**Culture conditions:** A design of a bioreactor usually ensures cell growth without cell damage, however, problems can arise from the complex rheological characteristics of plant cell suspensions since they become non-Newtonian fluids when they are growing and consequently, viscosity rises. In addition, possible foam formation bursting bubbles and shear stress can reduce the production. As is said before pneumatically driven bioreactors have a serious problem of foaming consequence of his bubble columns, this kind of bioreactors only mixes with bubbles that give a huge amount of problems such as non-homogeneous mixing, and shear stress with bubble rising and bursting. Plant cell cultures have plenty advantages in pharmaceutical industries and in cosmetic ones too, not only have advantage over traditional wild-plant harvesting, they are over bacteria cultures due their post-transcriptional capacity better than insect cells because of their well-known metabolic pathways and superior than mammal cells due to their safety ( no human viruses can sprout) and low cost.

As is said scientists well-know about their mediums, cultures, characteristics, and critical points of culture, modes and how to transform cells to more efficient ones. Now the main problem is the scale-up of our bioreactors, due the simplicity, low cost and safety pharmaceutical and cosmetic industries have an eye on this type of manufacture of new products, so bioreactors have to be improved.

After all the bibliographic research and thanks to summary tables of point 2.4, it is possible to determine the pros and cons of each bioreactor quickly. In the early past, industries uses stirred bioreactors, hollow fibers and bubble columns for their production as it's possible to see in Phyton Biotech™, which uses bioreactors of stainless steel. But now, single-use bioreactors are gone to set aside this last devices, PhytoCellTec™ (MibelleGroup) uses *Malus Domestica* for obtaining a component for a liposomal active ingredient based on stem cells from the Uttwiler Spätlauber apple, this cosmetic industry uses biowave bioreactors in sequence instead of huge stainless steel bioreactors. Single-use bioreactors are a reality in R+D laboratories due to their high amount of data in less time, low cost and safety but in industry stainless steel bioreactors are yet implemented; however low-cost safety, short time of production( it's not necessary to get clean and disinfected), high flexibility and simplicity are going to change this trend.

Products as their culture types are different, consequently the disposable bioreactors are lots, it seems that stirred bioreactors with different impellers are the most used kind of device, it has plenty of advantages and low cons, so the combination of stirred devices with bubble columns or Airlifts, suggests a great solution. Orbital bioreactors have more advantages than stirred but they only can be used for low-oxygen cell lines. Hydraulic and hollow fibers are only focused in immobilized and attached cells.

Root culture, due their difficulty and own characteristics has specific bioreactors, now the most used is the Mist bioreactor, Wilson bioreactor have a great capacity and product titer but it has plenty of contamination.

In the other hand, single use bioreactors, stirred ones have importance, but their lack in no foam formation and shear stress improves the acceptance of wave-mixing

Bioreactors, the only problem of this kind of device is the rheological issues of plant cultures, but this could be avoided by rise the rocking angle and velocity.

Finally, the hypothesis of the most suitable bioreactor for By-2 it was not confirmed nor rejected due the lack of data, nothing has yet published on the cultivation of BY-2 suspensions cells in commercial available orbitally, bag or disposable bioreactors. But using summary table 3.8, it seems that stirred bioreactors have more fresh weight and high pcv, using different inoculums and starter conditions (only differs in starting volume). Furthermore it could we assumed that orbitally shake are better than shaking ones, this is owing to the mechanical stress provided by the impact with plastic/glass of the bioreactor. With my personal experience with this kind of devices, they are not very trustable, the pcv, fresh weight can differ in large proportion, so it is necessary a well-stablished protocols and experience.

# Chapter - 15

# Entrepreneurship in Plant Tissue Culture

The capacity and willingness to develop, organize and manage a business venture like plant tissue culture is always challenging in order to make a profit. Entrepreneurship is generally combined with land, labor, natural resources and capital can produce profit. Entrepreneurial spirit is characterized by innovation and risk-taking, and is an essential part of a nation's ability to succeed in an ever changing and increasingly competitive global marketplace. The designing of PTC certain elements is essential for a successful operation. More than anything a solid knowledge about the subject and required technology are essential. The correct design of a laboratory will not only help maintain asepsis, but it will also achieve a high standard of work. Careful planning is an important first step when considering the size and location of a laboratory. It is recommended that visits be made to several other facilities to view their arrangement and operation. A small lab should be set up first until the proper techniques and markets are developed. A convenient location for a small lab is a room or part of the basement of a house, a garage, a remodeled office or a room in the headhouse. The minimum area required for media preparation, transfer and primary growth shelves is about 150 sq ft. Walls may have to be installed to separate different areas. Once the business picks up and demand increases then one can think of expanding the lab based on the demand. Larger labs are frequently built as free-standing buildings. Although more expensive to build, the added isolation form adjacent activities will keep the laboratory cleaner. Prefabricated buildings make convenient low-cost laboratories. They are readily available in many sizes in most parts of the country. Laboratory requirements and techniques are in length described in the earlier chapters Built-in-place frame buildings can also be used. Consideration should be given to the following:

1. Check with local authorities about zoning and building permits.

2. Locate the building away from sources of contamination such as a gravel driveway or parking lot, soil mixing area, shipping dock, pesticide storage, or dust and chemicals from fields.

3. A clear span building allows for a flexible arrangement of walls.

4. The floor should be concrete or capable of carrying 50 pounds per square foot.

5. Walls and ceiling should be insulated to at least R-15 and be covered inside with a water-resistant material.

6. Windows, if desired, may be placed wherever convenient in the media preparation and glassware washing rooms.

7. The heating system should be capable of maintaining a room temperature at 25-degree C.

8. A minimum 3/4 in. water service is needed.

9. Connection to a septic system or sanitary sewer should be provided.

10. Electric service capacity for equipment, lights and future expansion should be calculated.

Biotechnology has been globally accepted as one of the important tools for direct application in agriculture. It has a strong and positive influence on the agricultural sector worldwide. Agricultural biotechnology includes PTC, applied microbiology, and applied molecular biology contributing to the production of crops with improved food, feed, fiber and fuel. The technique of PTC is well translated from 'concept' to 'commercialization'. Through PTC large number of true to the type plants could be propagated within a short time and space and that too throughout the year. For example, it may be possible to propagate 2-4 lakhs of Tissue Cultured Plants (TCP) from a single bush of rose against 10 to 15 plants by vegetative means. Also, it may take about 2-4 months to produce healthy planting materials by tissue culture means, whereas a minimum of 6-8 months is required for most species by the latest method of vegetative propagation. As an industry, PTC is no more a nascent industry in India. It is flourishing with multidirectional growth and multimillion dollar turn over. Several plants like, anthuriums, bananas, strawberries sugarcane, orchids, teak, sandalwood etc. are routinely propagated by tissue culture technique and are being traded domestically and internationally for nearly three decades. Since PTC is a powerful technique for mass production in many crops, it has become an important tool in the nursery and farming industry. PTC technique has been responsible for bringing about the second green revolution in our country. The growth of PTC industry in India, its impact on the growing needs of the market, its business potential and the challenges this industry is facing are the matters of serious concern. Micropropagation is the application of tissue culture technique to the propagation of plants starting with very small parts grown aseptically in a test tube or other suitable containers. Micropropagation is one of the key tools of plant biotechnology that has been extensively exploited to meet the growing demands for elite planting material in the current century. There exists a large demand for disease free clones of superior quality plants in ornamental, horticultural, floricultural and agro-forestry sectors, which form the core sectors of agriculture. This need has been successfully tapped through micropropagation by the application of techniques of plant tissue culture thereby effectively translating the concept of technology for the commercial needs. As a result, several hundred plant tissue culture laboratories have come up worldwide, and more so in India.

In India, there are about 225 large commercial plant tissue culture units with a minimum production capacity of about 1 million plants per year from each of the units. Among these, at least 20 of the units have larger production capacities, with 5 to 10 million plants/year. In addition, there are hundreds of smaller units with 0.2 to 0.5 million plant production capacities where single crops are being produced. The Government of India has identified micropropagation industry as a priority area for further research, development and commercialization. Over the last 20 years, the Ministry of Science and Technology has supported 150 projects for research and development in this field. The favorable policies from the Ministries of Science and Technology, Commerce, Industries and Agriculture, Government of India have encouraged entrepreneurs and technocrats to set up more than 50 commercial units between 1987 and 1995 with a total installed capacity of about 210 million plants per annum. From 1986 to 1989 the targets achieved were 50% of the installed capacity. In 1991, there was a decline and only 20% of the target was achieved. In 1996, there was an increase in the number of plant tissue culture units and as a result most of the units had to suffer under-utilization of their facilities. The history of commercialization of plant tissue culture in India is a story of 'Rise and Fall and Rise Again'. The percentage increase in production decreased by 50% from 1991 to 1994 and in 1998 there was a negative growth showing rapid decline. However, between 1999 till date there has been an average of 35% rise in tissue culture production per year. This trend resulted in better capacity utilization of the existing facilities by 2022 and additional facilities are now being set up to increase the total installed capacity in the country to several million plants per annum.

## Growing demand and concerns

The demand for micro-propagated plants in agriculture, horticulture and in social forestry is growing by the day, since the traditional methods of propagation do not yield sufficient quantity and in some crops, they are cumbersome. Along with the demand the concern for developing quality plants is increasing. The emerging scenario on the growing use of tissue culture plantlets predicts that each state in our country should, at least, have ten tissue culture laboratories. The major consumers of tissue culture plants are the State Agriculture and Horticulture Departments, Agri Export Zones (AEZs), sugar and paper industries, private farmers and floriculturists. State-wise, the requirement of the crop type is different for the domestic consumption. It is important to note that the demand for some crops like banana, grapes, pineapple, strawberry, sugarcane, potato, turmeric, ginger, cardamom, vanilla and ornamentals like anthuriums, orchids, chrysanthemums, rose, lily, and gerberas are on the rise in different states in the country. Small quantities of medicinal plants like Aloe, Coleus, Chlorophytum, Digitalis, Melaleuca, Patchouli, Gloriosa and forestry crops like Teak, Rosewood, Sandal wood, Bamboo, Eucalyptus, Mangium are also produced and consumed in the domestic market. In 2007-08 the overall production of tissue culture plants was 145 million plants of the above species valuing Rs. 136 crores, with a growth rate of 20-25%. The consumption of plants has been approximately 450 million plants with banana constituting 41% share followed by sugarcane at 31% and ornamentals at 14%, spices at 6% and medicinal plants at 4%. The growth in demand for tissue culture banana has increased at a high rate of 25-30% and a similar trend for other crops is observed particularly for that of sugarcane due to the introduction of ethanol blended petrol. It can be noted that there is growing awareness of superiority of tissue cultured plants, and demand for crops like banana grapes, papaya, ginger, turmeric, cardamom, vanilla, potato, Jatropha is increasing When it comes to the international demand, the foliages and ornamentals have a great potential and the products have an unending elongated list. Major pot plants and landscaping ornamentals like Ficus, Spathiphyllums, Syngoniums, Philodendrons, Nerium, Alpenia, Yucca, Cordylines, Pulcherrima, Sansevieria, Gerbera, Anthuriums Rose, Statis, Lilies, Alstromeria etc. are routinely produced by various plant tissue culture laboratories in India. In2008 about 212.5 million plants including 157 million ornamental plants amounting to 78% of the total production are reported. There has been almost 100 % increase in the demand and production of PTC grown plants by 2019. It may be pointed out that tissue culture laboratory can also be used to produce biofertilisers like rhizobium, azotobacter, azospirillum, phosphate solubilising bacteria culture as well as mushroom spawn culture that indirectly contribute to the agricultural sector. The aggregate production capacity of the established commercial tissue culture units is estimated at 3 billion plants per annum. It is important to note that the installed production capacity is not always utilized completely and only 80-85% utilization is generally best utilized. The demand projections of about 4 billion plants requirement in the domestic market are highly conservative by 2022. Assuming that only 50% of the installed capacity is being used for addressing the domestic demand, the remaining capacity is used to cater to the export demand; still there is a large gap between the demand and supply. This clearly indicates a need for setting up additional units and supply plants with more competitive prices for improving the agricultural productivity, and enhancing the social status of the farmers.

## Government's Efforts

To encourage the tissue culture industry, various central and state government departments have framed several schemes and have announced incentives.

**Ministry of Agriculture:** The Department of Agriculture and Cooperation under the Ministry of Agriculture, Government of India provides financial assistance up to Rs. 21 lakhs and Rs. 10 Lakhs for setting up tissue culture units in public and private sectors respectively, subject to a maximum of 20% of the project cost. Under integrated development of fruits scheme, financial assistance in the form of subsidy, up to 50% is provided for purchase of tissue culture banana plants by various state Governments. The Government of India has set up a national facility for virus diagnosis and quality control of tissue culture plants at New Delhi with 5 satellite centers catering to the needs of the tissue culture industries in various parts of the country.

a. **Agriculture and Processed Food Products Export Development Authority (APEDA):** Under the Ministry of Commerce and Industry, state-of-the-art airfreight trans-shipment centre has been set up for tissue culture plants (perishables) at New Delhi, Bombay and Bangalore airports. Airfreight subsidy up to 25% of the freight cost is provided to tissue culture plants. 50% subsidy is given for the development of infrastructure like refrigerated van, packing, export promotion, market development, consultancy services, feasibility studies, organization building and human resource development. Financial assistance is also given for strengthening quality control facilities and implementation of ISO 9000.

b. **National Horticulture Board (NHB):** For setting up tissue culture lab there is a provision for back-ended capital subsidy not exceeding 20% of the project cost with a maximum of Rs. 25 lakh per project. Such subsidies are also extended to build up greenhouse and climate-controlled poly house/shade house.

c. **Small Farmers Agri-business Consortium (SFAC):** SFAC under the Ministry of Agriculture gives soft loans up to 50 lakhs for setting up small tissue culture labs by co-operative societies formed by small scale farmers.

d. **Department of Biotechnology (DBT):** DBT supports research and development projects across the country at various laboratories in the universities and research institutions for development and standardization of tissue culture protocols. The private tissue culture units are entitled for expansion of existing units as a Phase II activity under a scheme called Small Business Innovation Research Initiative (SBIRI). To promote the adoption of tissue culture technology by the industry and the end user, the department has established two micropropagation technology parks (MTPs) which provide a large number of service packages and have an important mandate of training and generating skilled manpower. The MTPs have transferred about 10 technologies to the industry and have also provided consultancy and taken up turn-key projects for various end users and state departments. The department has also set up a national facility for virus diagnosis and quality control of tissue culture raised plants, which are located at 6 different centers in India to ensure supply of disease-free plants to the end users.

e. **State level incentives:** The states of Karnataka, Gujarat, Maharashtra, and Andhra Pradesh are giving financial assistance for setting up tissue culture units under the new agro-industrial policy. Karnataka gives capital subsidy of 20% on investments.

All the above schemes have encouraged the establishment of tissue culture industry, which in turn have tremendously improved the demand for tissue culture generated quality planting material. A concerted effort is being made by the Government and the Industries to ensure that plant tissue culture, a technology with enormous commercial potential, would be an important industrial activity during the $2^{st}$ century.

**Creation of New Varieties and Status of Micropropagation in India**

The Indian scenario of tissue culture industry clearly indicates that it is a flourishing industry with about 225 large scale tissue culture units with a total production capacity of nearly 3 billion plants per annum currently. The analysis of the product range indicates that it is mainly concentrating on ornamentals, and horticultural crops. Fruit plants like banana, strawberry, pineapple, papaya; vegetables like tomato and potato, spices like cardamom, turmeric, ginger, pepper; plantation crops like sugarcane, vanilla, tea, coffee, Jatropha, and a very few forest crops like Eucalyptus, Paulownia, Sandal and Teak are being produced. Among the Asian countries that are active in commercial tissue culture plants production, it is in India that about 225 commercial units are functioning, while Indonesia and Japan have 53 units each, Korea has 30 and Thailand has 28 units. Other global countries involved in tissue culture plants production include European countries, USA, Canada, Australia, New Zealand, Israel, Middle East, South and Central America and Africa. All these countries together produced about 9 billion plants in the year 2019.

The use of plant tissue culture to produce somaclonal variation is one means of generating variation that may be needed in breeding program. This is particularly true in species that are traditionally propagated asexually or for which only few cultivars are available. Deliberate attempts to induce variations in tissue culture have been in progress for the last 60 years and a large number of variants in ornamentals and horticultural crops have been reported. However, there are only a few instances where somaclonal variations have produced agriculturally desirable changes in the progeny. These include sugarcane - increase in cane and sugar yield, and resistance to eye-spot disease; potato - improvement of tuber shape, colour and uniformity, and late blight resistance; tomato - increased solids, resistance to *Fusarium*. In the ornamental sector, *Syngonium* provides an excellent example of somaclonal variant where 52 new cultivars, all somaclonal variants, were selected from large populations of tissue-cultured material grown in commercial greenhouses. All 52 cultivars can be traced back to the original 'White Butterfly' clones. Each variant remained stable enough to become a named cultivar. Several of the cultivars have pink or reddish coloration in the foliage (Infra-Red, Bronze, Roxane is a few new varieties) that was not evident in 'White Butterfly'. Similarly, Labland Biotech, Mysore has obtained one unique, stable variant of *Spathiphyllum* with golden yellow leaves that has not been found in any Spathiphyllum varieties so far. It is being registered as the new variety of Spathiphyllum named 'Sona' and is being multiplied commercially for both domestic and overseas market in large numbers.

Among the other agricultural crops, CIEN BTA-03, a variant of Williams's variety of Banana resistant to yellow Sigatoka disease; AT626 & BT 627 of sugarcane variant resistant to sugarcane mosaic virus A and B are released for commercial usage. In addition, 'Ono', a sugarcane variant from variety, Pindar resistant to Fiji disease; ATCC 40463, a tobacco variety with enhanced flavour; DK 671, corn variety with higher yield, with lasting green colour and higher seedling vigour are the varieties from overseas inventions. Bio-13 a variety of Citronella released by CIMAP, India; Bio-902, Bio-YSR variants of *Brassica* parent 'Varuna' with enhanced seed yields are from India that are being multiplied and are cultivated successfully on commercial scale.

The impact of tissue culture technology in bridging the gap between the demand and supply could be exemplified by banana and Jatropha. It is now almost certain that only this technology can help in catering to the needs of quality planting material of banana and Jatropha. Banana is being cultivated in India in an area of about 500,000 hectares with an average productivity of about 15 kg of yield per plant. However, by replacing the conventional methods of use of suckers with tissue cultured plantlets, the productivity can be enhanced to about 50 kg/plant from the same area. In countries like Australia and Central America, bananas are always cultivated with tissue cultured plants after indexing for virus. At present, India is the largest producer of banana in the world with about 30% of total global production. But the export market share is a meager 1%. With increased productivity/unit area, the export capabilities can certainly be improved. This is possible by adopting the cultivation with virus indexed, tissue cultured plants instead of using the conventional suckers. Similar is the case with *Jatropha curcas*. The demand within the country for quality Jatropha plants is about 5 billion. This huge quantum of quality planting material supply is possible either through the adoption of tissue culture technique or by providing hybrid seeds.

The main advantage of tissue culture technology lies in the production of high quality and uniform planting material that can be multiplied on a year-round basis under disease-free conditions, and supplied anywhere irrespective of the season and weather. However, the industry is technology-driven. This technology is the amalgamation of triple alliance: capital, labor and energy. Although, labor is cheap in many developing countries, the resources of trained personnel and equipment are often not readily available. In addition, energy, particularly electricity, and clean water are expensive. Acclimatization *en mass* is also another expensive part of the industry wherein sophisticated greenhouses are essential to generate suitable end products. It is necessary to have low-cost options for weaning/hardening of micropropagated plants and finally growing them in the field. The most important aspect of plant tissue culture industry is to handle the technology very carefully. The technology in the wrong hands or the wrong use of the technology leads to unproductive results

# Future Prospects of PTC Industry

Long-term agriculture and forestry could be sustainable, with the use of little or no crop-protection chemicals, low energy inputs and yet maintaining high yields, while producing quality material. Biotechnology-assisted plant breeding is an essential step to achieve these goals. Plant tissue culture techniques have a vast potential to produce plants of superior quality, but this potential has not been fully exploited in the developing countries. During their growth under *in vitro* conditions, plants can also be primed for optimal performance after transfer to soil. In most cases, tissue-cultured plants out-perform those propagated conventionally. Thus, *in vitro* culture has a unique role in sustainable and competitive agriculture and forestry. It has been successfully applied in plant breeding, and for the rapid introduction of improved plants. Bringing new improved varieties to market can take several years if the multiplication rate is slow. For example, it may take a lily breeder 15 to 20 years to produce sufficient numbers of bulbs of a newly bred cultivar before it can be marketed. *In vitro* propagation can considerably speed up this process. Plant tissue culture has also become an integral part of plant breeding. For example, the development of pest- and disease-resistant plants through biotechnology depends on a tissue culture based genetic transformation. The improved resistance to diseases and pests enables growers to reduce or eliminate the application of chemicals. The potential of plant tissue culture in increasing agricultural production and generating rural employment is well recognized by both investors and policy makers in developing countries. These trends have augmented the firm footage of tissue culture industry as an established input into agriculture and have further opened up avenues for future growth. The plant tissue culture technology has been very successful as an industry and has greatly contributed to successful agriculture. The technology has created several employment opportunities and opened up many entrepreneurial fields. Usage of tissue culture-generated plants has increased productivity per unit area, particularly in horticultural crops. This industry has made available different unique commercial plant species such as ornamentals and foliages in large scale, which were not produced earlier by the conventional methods. Tissue culture has been one of the main technological tools and reasons that have contributed to the 'Second Green Revolution and Gene Revolution'. India is being looked upon by the world as the main technology base for production and supply of economically important plant varieties. With more innovative work, and intensive exploitation of our flora, the tissue culture technique will help us in consolidating our leadership at the global level.

## Production of Secondary Metabolites

Many thousands of chemicals are produced only in plants. Only few % of the world's plant have been scientifically named and only few compounds have been screened for the production of novel and useful compounds. Around 120 drugs are derived from plants. In western world around 25% of pharmaceuticals are derived from extraction of plants. These compounds are chemically complex and non-proteins, they have separate metabolic pathway. Due to less knowledge on the metabolic pathway, we couldn't enhance the metabolic products. In order to increase this production, bioprocess was introduced in plant cell cultures. Some of the plant products:- dyes, food colours, flavours, fragrances, insecticides and herbicides. The chemical compounds produced by plants are collectively referred to as phytochemicals. Biotechnologists have special interest in plant tissue culture for the large-scale production of commercially important compounds. These include pharmaceuticals, flavours, fragrances, cosmetics, food additives, feed stocks and antimicrobials. Most of these products are secondary metabolites— chemical compounds that do not participate in metabolism of plants. Thus, secondary metabolites are not directly needed by plants as they do not perform any physiological function (as is the case with primary metabolites such as amino acids, nucleic acids etc.). Although the native plants are capable of producing the secondary metabolites of commercial interest, tissue culture systems are preferred.

### Advantages of producing compounds from Plant Cell Culture

»   Control of supply of product independent of availability of plant itself and climatic, geographical and governmental restrictions etc.

»   High growth and turnover rate as compared to natural plant.

»   Reduction in time and space requirement for the production of desired chemicals.

»   Strain improvement with programs analogous to those used for microbial system.

### The production process comprises of seven aspects.

1.   Selection of cell lines for high yield of secondary metabolites

2.   Large scale cultivation of plant cells

3.   Medium composition and effect of nutrients

4. Elicitor-induced production of secondary metabolites

5. Effect of environmental factors

6. Biotransformation using plant cell cultures and

7. Secondary metabolite release and analysis.

## Limitations/Disadvantages

1. In general, in vitro production of secondary metabolites is lower when compared to intact plants.

2. Many a times, secondary metabolites are formed in differentiated tissues/organs. In such a case, culture cells which are non-differentiated can produce little.

3. Cultured cells are genetically unstable and may undergo mutation. The production of secondary metabolite may be drastically reduced, as the culture ages.

4. Vigorous stirring is necessary to prevent aggregation of cultured cells. This may often damage the cells.

5. Strict aseptic conditions have to be maintained during culture technique: Any infection to the culture adversely affects product formation.

The following four different culture systems are widely used:

1. Free-cell suspension culture

2. Immobilized cell culture

3. Two-phase system culture

4. Hairy root culture.

(For details about the above methods, refer to related notes)

# Chapter - 16

# Automation and Robotics in Plant Tissue Culture

Plant tissue culture is an advanced clonal propagation technique for producing high quality and disease-free plants in a short time period. With the increasing demand in agriculture, horticulture, and herbal products. Especially in the production of commercial valuable crops like, roses, anthuriums, banana-propagules, berries, potatoes, orchids and saplings of teak, sandalwood, rosewood etc. Plant tissue culture has become increasingly important in large-scale clonal propagation. But the problem with plant tissue culture is that, it's a process which is very tedious and labour intensive. Novel automated micropropagation systems have been developed to reduce the labour cost as well as to increase the throughput and reduce contamination risk by minimising human contact. The automated systems have improved the rate of production of plants with high accuracy and also lowered the cost of production. Micropropagation is beneficial to pharmaceutical industry as it guarantees for the standardised formulations in plants or their parts to be uniform both quantitatively and qualitatively. The benefits of plant tissue culture are extensive in the agricultural world in many respects, the first being that it allows for the production of a significant number of plants in a very short period. Plant tissue culture is also advantageous to growers because the overwhelming number of disease-free plants can be produced using the tissue collected from a single parent plant e a plant which itself remains unharmed in the tissue harvesting process. Propagation through tissue culture also eliminates the possibility of any interruption in the growing season because it can be carried out inside a carefully regulated, controlled environment.

Traditionally, the plantlets are grown in a container filled with sterile growing medium containing necessary nutrients, carbohydrates and growth regulators. Under a laminar flow hood, the operator removes the plantlets from the growing medium one by one and cut them into two or three smaller micro-shoots. They are then planted into a new container containing fresh medium and incubated in a controlled environment until they are ready for next transfer. Decades ago, in North America or Europe, labour was the first major contributor (up to 80%) towards the cost of tissue-cultured plants Labour cost still remains one of the major costs in tissue culture industry in the developed countries. The second major factor contributing to high cost is electricity (e.g., 15%). Lighting of shelves and cooling of culture room consumes the major portion (e.g., 85%) of electricity. The only way to further its application is by increasing the efficiency and quality of the output while reducing the relative labour cost. Since the micropropagation process is highly repetitive and tedious, it becomes a good candidate for automation. In addition to reducing the relative labour cost, other advantage of automating the micropropagation process is the reduction of the risk of contamination. Introduction of microbial contamination by operational technicians is one of the most serious problems of plant cell and tissue culture. In addition, more systematic and consistent production, and better control of the products can be achieved by automating the process of micropropagation. Plants are living organisms and since all plants are different, it is technically challenging to develop a general automated process.

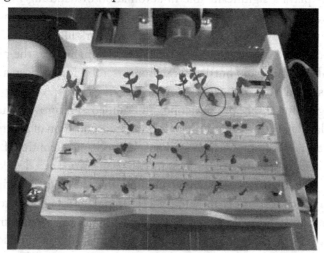

**Robotic Transplantation Unit**

In order to increase the speed of plant production and to avoid contaminations, automation of almost all steps of steps of plant tissue is the only way. Let us look into opportunities and strategies involved in the total automation of plant tissue culture industry. This is very much necessary and is the need of the hour now. The global plant tissue culture industry is expected to become a billion-dollar industry post

2025. It is growing at a rate of CAGR 8.5%. Flowering plants, wood yielding plants and horticultural plants are going to increase the demand for PTC grown plants and propagules. VIP-grade vegetable, especially the leafy vegetables are increasing becoming popular and they are invariably sourced from PTC grown plantlets.

Sterilization of the source plants to obtain the explants and further sterilization of the explants can be totally automated and it has been achieved by most of the PTC industries. Similarly, media preparation and dispensing of media into the vessels and sterilization of all vessels used are also completely automated. The process of high-speed inoculation with zero contamination and high accuracy is also partially automated. But this area requires lot of improvisation as it involves Vision-Manipulator-Controller Technology. The ease and flexibility achieved with the human eye-hand-brain combination cannot be easily simulated that easily. The next step in PTC is shifting of culture vessels into the growth room. This involves a simple technology and it has been achieved by many PTC industries already. The final step is trans-flasking, in which the PTC grown ex-agar plants are shifted to the pots with soil (sterile-soil). This step is very easy for automation.

Robotics plays a significant role in automation of handling materials in the industries. Service robots have also gained lot of attention in all almost all sectors. Robots need not always look like humans but they work like humans. In fact, they are meant for more efficiently and accurately than humans. For the production of agricultural, floricultural and horticultural products, robotics has been an extensively employed in several parts of the world for more than 35 years. A substantial amount of effort in agricultural robotics development is aimed at improving automation in operations related to the production, handing and transportation of plants. Robotics is divided into several sub-topics such as manipulator mechanism, control of mechanisms, end-effector design, sensing techniques, mobility, and work cell development. The major issues related to Robotics are economics and the man-machine interface Production of plants *in vitro* occurs at many levels, ranging from micropropagation for plant regeneration, large-scale production of field crops & forest trees, transplanting, handling, hardening packing and transportation. Commercial plant production systems have been developed in order to meet the market demand of the PTC grown plants In such instances, plant materials are frequently handled in large quantities. The need to use machinery in plant production to reduce human labor requirements, expand production competences, and improve consistency of plant products is predictable. The automation of plant production enables a small percentage of the population who are modern farmers to produce large quantities of flowers, vegetables, fruits and fibres on a desired schedule. Owing to the biological features of plants, they need to be altered in distinct ways, and sometimes with personal care. The cost of materials handling by manual labor and the availability of skilled personnel have become an increasing concern for plant producers. The dawn of computers brought about the opportunity of augmenting apparatuses and machines with artificial intelligence. Robots are programmable intelligent man-made machines furnished with sensory gadgets and changeable end-effectors, positioned and oriented by a mechanism, for executing manifold of activities and handling risky substances. In addition to mechanization functions, robots also provide various automation capabilities with flexibility, which can be of great value to plant production operations. In a broad sense, automation encompasses machine capabilities of information processing and task execution to facilitate a system's operation. Information processing includes the activities of information acquisition, organization, manipulation, interpretation, understanding, adoption, and presentation. Commonly applied information processing functions in an automated system are perception, reasoning, learning, and communication. Task execution requires task planning and mechanical work. Robotics is an integration of computer technologies, sensing and control techniques, and generic mechanisms. The result is a flexibly automated mechatronic system with in-built artificial intelligence is well suited for performing numerous materials handling maneuvers. Robots are normally working in concert with other sensing and materials handling devices within a defined space like the "Work-Cell". The types of devices included in a workcell determine its overall functionality. Careful planning and designing is a must at the systems level before the actual implementation of the "Work-Cell" design. Using of robotics in production of plants and plant products demands the integration of robot capabilities, plant culture, and the work environment. Robotics for the purpose of plant production includes the development and careful designing of manipulators, end-effectors, visual sensors, and travelling devices.

Large-scale clonal propagation of commercial crop plants demands certain cultural practices to be performed on the plants under specific environmental conditions. Some of the environmental conditions are mostly natural and some are modified or controlled. The mandatory cultural practices command the layout and materials flow of the production system. Both the cultural and environmental factors significantly affect when, where and how the plants are manipulated. In addition, the plants are expected to change their shape and size during growth and development. Individual plants within any given population will have significant variation in properties important to the robotic operation to be performed. The robots which are to be used for performing plant cultural tasks must recognize and understand the physical properties of each unique object and must be able to work under various environmental conditions in fields or controlled environments. They hence, require highly sensitive sensors which can work under the variable conditions as well as specialized manipulators and end-effectors. The environmental conditions are occasionally so severe with regard to high temperature, humidity, dust and/or rain that electrical circuit and material corrosion problems can be major concerns. These conditions must be considered when designing or selecting plant production robot systems. If the work object is not easily positioned in front of the robot, a travelling device is required. Several cultural practices are commonly known in the plant production industry. The ones which have been the subject of robotics research include division and transfer of plant materials in micropropagation, transplanting of seedlings, sticking of cuttings, grafting, pruning, and harvesting of fruit and vegetables.

Key Robotic component is the "Manipulator" The basic mechanism of a manipulator is defined by its degrees of freedom, the type of joint, link length, and offset length. Any kind of manipulator may be used, if its work envelope includes the position of the work object and the work efficiency is not a major concern. This is because the main task of a manipulator is to move and orient an end-effector to a position where it can interact with the work object. If a manipulator whose basic mechanism is not optimized for a particular production system is used, the work speed may be slow and the manipulator may only be able to place the end-effector at a singular point. This can potentially present some problems in a plant production system. The manipulator may have to risk a collision with the objects within the work envelope. Therefore, a mechanism optimized for the specific task is often required for a plant production robot. On the other hand, however, a manipulator which has a mechanism developed based on a specific operation is likely to have less flexibility in adapting to other operations. Nevertheless, a special purpose manipulator can still be used in performing various jobs by using different end-effectors. The factors normally considered in determining basic mechanism requirements for a manipulator of a plant production robot will be; Work envelope, Measure of manipulatability, Posture diversity, End-effector, Visual sensor, Discrimination, Distance measurement, and Travelling device.

Work envelope is normally used to reduce the need of moving the manipulator. Objects of work such as plants with fruit to be harvested or trays of plants to be worked on can be presented to a robot in the similar way. In such a case, many types of manipulator mechanism may be used; however, an appropriate spatial range for the positions of the work objects should be included within the work envelope. Manipulator is evaluated by the measure of manipulatability which implies easiness to move the end of the manipulator. The manipulator can have the possibility to avoid an obstacle before approaching the target object if necessary. The posture diversity is defined by the angle at the redundant space. A specific mechanism for an end-effector depends on a specific work object and an operation to be performed. Since the properties of the work object the robot directly handles and the type of operation are different from any others this should be unique. Most of objects for plant production systems have various sizes and uncertain shapes, at least to some degree, even if they are same varieties. A visual sensor is a very important external sensor of a robot just like the eyes for humans. The three important functions of visual sensors for a plant production robot are: discrimination, recognition, and distance measurement.

Each component for a plant production robot must be developed based on the physical properties of its work object. The horticultural aspects of the production system must also be considered to realize the practical use of various kinds of robots. A number of robotic based systems for multiplication of plant tissues and transplanting of plantlets have been developed by commercial companies and research institutions have been introduced.

From the late 1980s to the mid-90s, there was a global interest among different countries (particularly USA, UK, the Netherlands, Germany, Japan, and Australia) to develop automated technologies for micropropagation. Researchers and companies attempted in developing automated transplanting systems for propagating plants through tissue culture. For example, several companies in Japan incorporated automation systems in their tissue culturing fields. The Toshiba plant tissue culture system (Japan) was built for propagating nodules plants that consist of multiple meristems. The system involved two cooperating robots: a sensing robot and a cutting robot. A 6 degrees-of-freedom (DOF) robot picked up a plant from a tray and held it, then another unit sensed the correct position of the plant with a laser beam. A node was detected when the diameter of the stem exceeded a predetermined value, i.e. where the stem branches. A second 6-DOF robotic arm equipped with a tweezer-like gripper and scissor-like cutter cut the plant and transplanted it. The gripping force was measured with strain gauges on the gripper and was fed back to the motor to achieve a gentle gripping that did not damage the plantlet segment. The time required for recognition, cutting, and transplanting of a single node was 15 seconds. Another research group in Japan developed robotic systems integrated with 3D vision for plant cloning. Two robotic systems were developed for two different kinds of plants and each system employed separate vision systems for each of their cutting, adjusting and planting stages. Some research groups in Europe also developed automated micropropagation systems that utilised a laser cutter as the dissecting tool. Furthermore, a group of researchers in Germany proposed an automatic micropropagation system for grass. The system contained three modules, which were connected through an assembly belt. The plants being propagated came from a culture room and arrived at the first module, where they were pulled out by an end effector and placed on a pallet. The pallet then travelled to the second module where the plants were inspected by a vision system and cut by a knife. After that, the pallet moved to the third module, where the shoots were picked up and planted into a new culture vessel. Another group of researchers from a company in Australia developed an automated micropropagation system for timber and pulp. The system contained three compartments and pneumatic power was used to operate the system. In the first compartment, the trays with fresh growing medium and the harvest plants were loaded onto the system. The trays were then passed onto the second compartment, where the lids that used to cover the trays were removed and the plants were inspected by a CCD camera. Next, the plants were gripped and cut in the same compartment and the shoots were vacuumed to a funnel to be planted. Once the tray was full, it was passed to the third compartment to be collected. Way back in 1985 itself, Sevila studied an alternative grape vine pruning method to facilitate robotic applications. The new pruning method was evaluated and revised by a computer model and a laboratory scale robotic system was constructed. The system consisted of a cutting saw attached to a cutting arm, an image acquisition system, and an electronic controller. It was concluded that the method was physiologically and agronomically feasible. The robotic system was found workable with the new pruning method. In 1990, Simonton developed a robotic workcell for geranium stock processing. In 1992, Okamoto et al developed a tissue proliferation robot for dividing and transferring callus. A special end-effector equipped with a disinfected pair of razor blades and a suction pipe was carried by an articulated robot to perform the task. The robot was guided by a machine vision system in locating plant tissues on a petri dish. The average cycle time for dividing and transfer a callus was approximately 40 seconds. There have been a series of projects in developing robotic systems for transplanting seedling plugs and cuttings. Transplants are used in both greenhouse and field production of floral and vegetable crops because of their many advantages including uniformity, earliness, etc. The additional labor required in handling transplants, as opposed to direct seeding, is the transplanting operation. Geranium cuttings were processed using this system for vegetative propagation. A grafting robot was developed by Suzuki et al in 1993 for preparing cucumber seedlings. The grafting operation involved the preparation of scions and fixing/adhering the scions on stocks. It was reported that the cycle time for producing a grafted seedling was about 3 seconds. In 1993, Cornell University scientists worked on the development of a robotic grape pruner. A computer model of a vision guided pruner was developed by Ochs and Gunkel to study the effects of the robot components and the variation of vine and terrain on the pruning accuracy. The design of a digital regulator and tracking controller for a robotic electro-hydraulic pruner was discussed by Lee et al in 1994. In Okayama University researchers developed a robot which would perform a number of operations in vineyard by Monta et al in 1994. Yamada et al also developed a fully automated robot-based grafting robot in 1995.

In Taiwan, a research group developed a system that achieved 78.2% success rate in grasping the plantlet in the desired position to obtain higher success rate in preparing explants (Huang & Lee, 2010). Two CCD cameras were used to create a 3D coordinate system and the information was passed to an industrial grade 6-DOF robot arm for grasping. As the automation technology advanced, an increased number of novel methods were developed and automated. These included the use of alternative techniques such as autotrophic tissue culture systems, in which plantlets are grown in large vessels within controlled environment, and microponic systems, in which plantlets are grown in nutrients solution in small scale. The authors stated that the production cost for automating the autotrophic systems is high due to the equipment needed for controlling the desired culture conditions. On the other hand, while the microponic systems do not require such equipment to control the environment, the micropropagation process should be carried out under clean condition. All the above-mentioned systems were rather complicated with a relatively large infrastructure, specifically designed for a particular method, and less efficient due to the fact that the plantlets were processed and dissected one by one. Recently, the automation of transplanting systems has become more mature technologically. Some companies have implemented automation in the transplanting stage but are limited in automating the cutting and prepping of the plantlets. A group of researchers in China proposed an automated micropropagation system to cut and transplant potatoes in batch processing. Two assembly belts were used to transfer the jars with the plantlets and empty jars filled with growing media, respectively. At the entrance of the system, a robotic manipulator was used to hold the jar that contains the plantlets and rotated it to the desired orientation. After that, this jar was transported to a grasper, which was used to pull all the plantlets out of the bottle at once. Then, the manipulator inverted this jar and transported it to a gatherer, which was used to collect all the plantlets into a bundle while a cutter was used to cut the plantlets into small shoots. The shoots then fell into the empty jars transported by another assembly belt. Nevertheless, the success rate of grasping the plantlets out of the jar was around 85.6% and the surviving rate for the successfully cut shoots was around 90.6% because not all the shoots properly fell into the growing media. Even though the implementation of automating micropropagation process has been around for decades, they are either inefficient or expensive to be implemented because of its sophisticated features; thus, the automated systems have not been commercialised widely. Another limiting factor was that every plant varies in different ways so the automated systems designed for specific category of plants may not work properly with different species. In the current study, we propose a novel automated micropropagation system for plants that propagate through nodal cutting, such as blueberry plants. Different from previous works, uniquely designed culture vessels are proposed to provide easy access to the plantlets and process them in a batch in the upright position rather than taking the shoots out and process them one by one. In addition, the proposed system, which can be fit under a laminar flow hood for sterilisation purposes, is self-contained and small in size and is easier to be manufactured and commercialised.

Lee et al from Simon Fraser University, developed an automated micropropagation system is to increase the production rate of 'Biloxi' blueberry plantlets (*Vaccinium corymbosum*). In order to increase the production of the Biloxi plantlets, three novel design concepts have been implemented. The first novel design concept is the culture vessels, which have been redesigned to provide easy access to the plantlets. In the field of tissue culture, use of baby-food jars, autoclavable polypropylene or polycarbonate containers such as Magenta boxes, are a common practise. All of these vessels as well as the current method (i.e., the baby food jars) used at the Plant Tissue Culture lab, JRT Nurseries, have one common problem, which is the shoots may get easily entangled due to their nondiscrete placements inside the round jars. This may result in a non-uniform plant production. Another problem is that the storage space is not properly utilised due to the empty space between the adjacent jars. Different from the traditional storage method (e.g., jars), this system utilises a tray-stacking method for storage. In the preliminary design of Lee et al, the culture vessels are detachable. This also allows the vessels to be easily sterilised after each use. Each row of the vessel (named 'Detachable Vessel Vector' or DVV) will detach from the rest and each compartment of the DVV is dimensionally optimised to accept a single plantlet. In doing so, the access to the plantlets is increased and the planting area for storage is optimised. The jar is around 85 mm in diameter, which requires a shelf area of 7225 mm². When four DVVS are grouped together and enclosed using a transparent cover (with 1 mm wall thickness) they occupy a comparable shelf area of 7076 mm². The inner dimension of each DVV cell is 10X10 mm and the wall thickness is 2 mm, which is sufficient to prevent sink marks and warpage from injection moulding.

SmartClone™ has the potential, by improving tissue culture growth and increasing operational efficiency, to massively extend the benefits of cloning to crops currently depending on seed propagation, with applications in broad acre farming horticulture, forestry and biofuels. SmartClone™ represents a fast-track delivery system for genetic improvements obtained through breeding. NuPlant is committed to the improvement of clonal propagation techniques to produce tougher plants faster

NuPlant focuses on the commercialization of its proprietary SmartClone™ plantlet propagation systems through the sale of SmartClone™ robotic systems and related consumables to enable tissue culture operations worldwide to increase production and improve the quality of the plant propagation material produced. NuPlant also operates a tissue culture facility with annual robotic SmartClone™ throughput of, depending on the characteristics of specific plantlets produced, between 5 and 10 million plantlets per year. The self-contained machine capable of processing 12 explants per minute comprised the following:

» Dual camera stereo imagery and parallel processing of images for identification of plant nodes for automatic cutting

» 9 independently controlled linear and two rotary axes with stepper motor actuation

» 2 tray-dispensing magazines

» 3 independent 'smart location' control conveyors

» 3 independent harvesting, cutting and planting end effectors

» Ultrasonic and heat augmented tool sterility

» Air blast and vacuum waste removal

» Rugged construction for operation in a harsh nursery environment

This system many advantages such as, increased production rate by 500% for equivalent labor utilizations, eliminated difficult, highly skilled and labor-intensive operations, Reduced contamination and a total of 12 in-built machines.

The Weighing, Imaging and Watering Machine (WIWAM) is an imaging platform for high-throughput phenotyping of *Arabidopsis* plants. The robot is capable of handling a large number of plants simultaneously and measuring a variety of plant growth parameters at regular time intervals. The standard platform consists of a table with a capacity of 396 pots, and a robotic arm, which moves individual pots regularly to separate stations for imaging, weighing and watering (F. Phenotyping is performed by an image analysis script that extracts rosette growth features such as the compactness, perimeter and stockiness of each plant automatically. The image sequences allow constructing reliable growth curves over time. WIWAM replaces labor-intensive manual work, saving time and costs. Because the growth conditions are standardized and larger sets of plants can be screened, the robot also allows obtaining statistically more significant results. In addition, the image analysis provides information which are overlooked by the naked eye. In our department the WIWAM is mainly used to investigate rosette growth under mild drought conditions, but it can be applied to a whole range of biological questions. Currently, we are looking into the possibility to commercialize this plant phenotyping robot. Furthermore, the WIWAM setup can be tailored to meet your specific needs.

In order to monitor growth parameters of *Arabidopsis* plants *in vitro*, the In Vitro Growth Imaging System (IGIS) was built. The basic setup consists of a rotating metal disk, which can accommodate up to ten Petri dishes and a single-lens reflex camera, which captures a top view picture of the germinating seedlings. The disk is put into motion by a step motor driving a central axle. The motor is connected to a programmable logic controller (PLC), allowing the integration of timers and sensors that control the motor. To prevent condensation at the top of the Petri dishes, the disk has to be cooled. Therefore, the system is connected to an air cooling system, propelling cooled air into a metal ring underneath the disk. The phenotyping platform can be used under continuous light conditions, but to enable viewing the plants during the dark period, the platform is equipped with infrared (IR) light emitting diodes (LEDs) with a wavelength peak around 940 nm, which is well outside the visible spectrum for plants. Furthermore, to allow the reflex camera to capture the IR light and to ensure that the pictures taken during the light period are comparable to those taken during the dark period, some adjustments are employed. IGIS is placed in the tissue culture rooms to ensure controlled growth conditions.

**Automatic Plant Tissue Culture System**

During a growth experiment, the system runs for a period of 21 days with the specifications to move one position every six minutes, bringing each of the ten Petri dishes underneath the camera position within the time frame of one hour. For each Petri dish, twelve seeds are evenly spread over the plate, to prevent plants overlapping at the final stages, meaning that the individual rosette growth of 120 plants was followed over 480 hours in a typical time lapse sequence. An automated image analysis and visualization pipeline was created for IGIS. First, positions of the plants to be analyzed are marked in a so-called mask file. The selected plants on each plate are numbered automatically based on the position of the corresponding disks in the mask images (F. Then, individual plants are extracted and rosette parameters are measured. These values are extracted for all plants on every plate at each time point and are used for further calculation of mean rosette area, compactness values and other time-derived parameters like relative rosette growth rate and relative change in rosette compactness. Finally, the data is automatically plotted in a number of graphs, revealing temporal rosette growth patterns of *in vitro*-grown *Arabidopsis* plants. Another phenotyping tool, the Multi-camera In vivo Rosette Growth Imaging System (MIRGIS), was developed in the Systems Biology and Yield group. It is well known that the physical environment affects plant phenotypes immensely and growing *Arabidopsis* plants in soil can lead to increased biological variation within identical genotypes. This new platform allows automatic tracking of individual plants after germination. The system consists of several single-lens reflex cameras and a laptop. Three cameras are mounted above six trays of *Arabidopsis* plants in a custom designed rack together with fluorescent lamps and placed in a controlled growth chamber (**Figure 5**). An in-house computer script was developed so that each camera would take a picture of the two trays underneath on a daily basis. The setup can image a total of 144 plants, per three cameras. The basic concept is that four seeds are sown per pot and followed from early development. At a specific point during development, a seedling selection is done in order to reduce the variation within the pool of in soil-grown plants, which mainly arises by differences in germination time. This is performed before the plants start to overlap. Therefore, the projected areas of all seedlings of a specific genotype (which are usually spread over the different trays) are measured and the median seedling area of that genotype is calculated. Then, the seedling with an area closest to the median value is selected within each pot of that genotype. A selection image is generated in which the selected plant is indicated by a red circle. This image is used to manually remove the remaining plants in the pots. This methodology ensures that the seedling that is selected per pot is the one closest to the average plant in the population of its genotype. The location of the selected seedling is afterwards used to extract the plants throughout the complete image sequence, allowing for the construction of growth curves and the calculation of relative growth rates. This phenotyping setup is designed for *Arabidopsis* rosette growth analysis.

The maize WIWAM, also called 'SHRIMPY' (System for High-throughput Recording and Imaging of Maize Phenotypes related to Yield), is an imaging robot designed for crops in the Systems Biology of Yield Group of Professor Dirk Inzé. The system has a capacity of 156 plants. Like the *Arabidopsis* WIWAM, it is located in a controlled-environment growth chamber and provides for automated imaging and automated weighing and irrigation of plants according to a preset scheme, specific for each plant or group of plants. This may include fixed quantities of solution, or irrigation to a certain target weight determined by the requested soil humidity levels. The system is thus ideally suited to impose soil water deficit treatments. In contrast to the *Arabidopsis* WIWAM, plants are positioned in tables which are sorted in rows. The robot arm pushes the tables aside to create the required space in between the rows for the robot arm to locate a pot and lift it out of the table. The robot then takes it to the rotating platform of the weighing and irrigation station. For the imaging of plants, pots are taken to a designated area at the back of the growth chamber where plants are rotated in front of a RGB camera for multiple angle imaging. The acquired images are used for the three-dimensional reconstruction of plants and the extraction of quantitative growth-related traits. Experimental setups, experiment metadata and results are managed by PIPPA (PSB Interface for Plant Phenotype Analysis), the central user interface and database.

PHENOVISION is a greenhouse infrastructure for automated, high-throughput phenotyping of crops with a capacity of 392 plants. Pots are transported in carriers on a conveyor belt system. Both pots and carriers have unique identifiers , which makes it possible to treat each plant individually in transit from its position in the stationary growth area of the system to the weighing and irrigation stations and the imaging cabins.

Currently, the system includes three weighing and irrigation stations with rotating platforms and the possibility to apply water and up to three different solutions. Soil water or nutrient deficit conditions can thus be imposed on plants.

The imaging cabins are enclosed areas with camera-adapted lighting conditions and a lift with a rotating platform. At present, three camera systems are available in the cabins. The first one consists of RGB cameras in a multi-view imaging setup for the three-dimensional reconstruction of plants and the measurement of growth-related phenotypic traits. Plant physiology-related traits are measured or approximated by exploring a larger stretch of the electromagnetic spectrum. A thermal infrared camera captures energy emitted at 8-13 μm. The corresponding contextual plant and leaf temperature is used as a proxy for plant water use behavior. A state-of-the-art hyperspectral imaging system, consisting of a visible to near-infrared camera (VNIR, 400-1000 nm) and a short-wave infrared camera (SWIR, 1000-2500 nm), constitutes a novel tool for close-range sensing of plant physiological traits based on reflectance spectra captured on whole-plant and individual leaf level.

Smart features of the infrastructure include a 'handling zone' where the system can bring and retrieve a requested number of plants belonging for example to a certain genotype or treatment. As the handling zone is accessible by users of the system, it allows for visual observations of plants or manual actions on plants, such as the measurement of specific plant traits or the extraction of plant parts for molecular or biochemical analyses. A second smart feature of the system is the possibility to load external plants into the system, for example plants grown in another greenhouse compartment or growth chamber, in order to have them imaged and/or treated at the weighing and irrigations stations.

Environmental parameters, including air temperature, relative humidity and light intensity (photosynthetically active radiation), are continuously monitored in the greenhouse to direct the greenhouse heating, ventilation, humidification and lighting system, but also to support genotype-environment interaction studies in greenhouse conditions.

Experimental setups, experiment metadata and results are managed by PIPPA (PSB Interface for Plant Phenotype Analysis), the central user interface and database.

PHENOVISION has been developed and constructed in collaboration with SMO bvba and is financially supported by a grant of the Hercules Foundation (Belgium) awarded to Professor Dirk Inzé.

# References

Ahloowalia BS, Prakash J, Savangikar VA, Savangikar C (2003) Plant tissue culture. Low cost options for tissue culture technology in developing countries. In: Proceedings of a technical meeting organized by the Joint FAO/IAEA Division of Nuclear Techniques in Food and Agriculture and held in Vienna, 26–30 August 2002 (pp. 3–10). Vienna, Austria: International Atomic Energy Agency. ISBN 92–0–115903–X

Ahloowalia BS, Savangikar VA (2003) Low-cost options for energy and labour. Low-cost options for tissue culture technology in developing countries. In: Proceedings of a technical meeting organized by the Joint FAO/IAEA Division of Nuclear Techniques in Food and Agriculture and held in Vienna, 26–30 August 2002 (pp. 41–46). Viennam Austria: International Atomic Energy Agency. ISBN 92–0–115903–X

Ali RM, Abbas HM. Response of salt stressed barley seedlings to phenylurea. Plant Soil Environ. 2003;49:158–162. doi: 10.17221/4107-PSE.

Bhattacharyya P, Kumaria S, Bose B, Paul P, Tandon P. Evaluation of genetic stability and analysis of phytomedicinal potential in micropropagated plants of *Rumex nepalensis*—a medicinally important source of pharmaceutical biomolecules. J Appl Res Med Aromat Plants. 2017;6:80–91. doi: 10.1016/j.jarmap.2017.02.003

Bhattecheryya P, Kumaria S, Diengdoh R, Tandon P. Genetic stability and phytochemical analysis of the in vitro regenerated plants of *Dendrobium nobile* Lindl., an endangered medicinal orchid. Meta Gene. 2014;2:489–504 doi: 10.1016/j.mgene.2014.06.003.

Binoy J, Silja PK, Dhanya B, Pillai B, Satheeshkumar K. In vitro cultivation of hairy roots of *Plumbago rosea* L. in a customized reaction kettle for the production of plumbagin—an anticancer compound. Ind Crops Prod. 2016;87:89–95 doi: 10.1016/j.indcrop.2016.04.023.

Borchardt JK. The beginnings of drug therapy: ancient Mesopotamian medicine Drug News Prespect. 2002;15:187–192. doi: 10.1358/dnp.2002.15.3.840015

Bose B, Kumaria S, Choudhury H, Tandon P. Assessment of genetic homogeneity and analysis of phytomedicinal potential in micropropagated plants of *Nardostachys jatamansi* a critically endangered, medicinal plant of alpine Himalayas. Plant Cell Tissue Organ Cult. 2016;2:331–349. doi: 10.1007/s11240-015-0897-x.

Buyel JF, Twyman RM, Fischer R. Very-large-scale production of antibodies in plants: the biologization of manufacturing. Biotechnol Adv. 2017;35:458–465. doi: 10.1016/j.biotechadv.2017.03.011.

Castilho A, Gattinger P, Grass J, Jez J, Pabst M, Altmann F, et al. N-Glycosylation engineering of plants for the biosynthesis of glycoproteins with bisected and branched complex N-glycans. Glycobiology. 2011;21:813–823. doi: 10.1093/glycob/cwr009.

Cetin ES. Induction of secondary metabolite production by UV-C radiation in *Vitis vinifera* L. Öküzgözü callus cultures. Biol Res. 2014;47:37. doi: 10.1186/0717-6287-47-37.

Chen C-C, Chang C-L, Agrawal DC, Wu C-R, Tsay H-S. In vitro propagation and analysis of secondary metabolites in *Glossogyne tenuifolia* (Hsiang-Ju)—a medicinal plant native to Taiwan. Bot Stud. 2014;55:45. doi: 10.1186/s40529-014-0045-7

Chung IM, Rekha K, Rajakumar G, Thiruvengadam M. Production of glucosinolates, phenolic compounds and associated gene expression profiles of hairy root cultures in turnip (*Brassica rapa* ssp. *rapa*) 3 Biotech. 2016;6:175. doi: 10.1007/s13205-016-0492-9.

Debnarh M, Malik CP, Baisen PS. Micropropagation:a tool for the production of high quality plant based medicines. Curr Pharm Biotechnol. 2006;7:33–49. doi: 10.2174/138920106775789638.

Fabricant DS, Farnsworth NR. The value of plants used in traditional medicine for drug discovery. Environ Health Perspect. 2001;109(supplement):69–75.

Fischer R, Vasilev N, Twyman RM, Schillberg S. High-value products from plants: the challenges of process optimization. Curr Opin Biotechnol. 2015;32:156–162. doi: 10.1016/j.copbio.2014.12.018.

Furusaki S, Takeda T. Bioreactors for plant cell culture, reference module in life sciences. Dordrecht: Elsevier; 2017.

George P, Manuel J. Low cost tissue culture technology for the regeneration of some economically important plants for developing countries. Int J Agric Environ Biotechnol. 2013;6:703–711.

Georgiev MI, Agostini E, Ludwig-Muller J, Xu J. Genetically transformed roots: from plant disease to biotechnological resource. Trends Biotechnol. 2012;30:528–537. doi: 10.1016/j.tibtech.2012.07.001.

Georgiev V, Ilieva M, Bley T, Pavlov A. Betalain production in plant in vitro systems. Acta Physiol Plant. 2008;30:581–593. doi: 10.1007/s11738-008-0170-6.

Govindaraju S, Arulselvi PI. Effect of cytokinin combined elicitors (1-phenylalanine, salicylic acid and chitosan) on in vitro propagation, secondary metabolites and molecular characterization of medicinal herb—*Coleus aromaticus* Benth (L) J Saudi Soc Agric Sci. 2016 doi: 10.1016/j.jssas.2016.11.001.

Grzegorczyk-Karolak I, Kuźma Ł, Skała E, Kiss A. Hairy root cultures of *Salvia viridis* L. for production of polyphenolic compounds. Ind Crops Prod. 2018;117:235–244. doi: 10.1016/j.indcrop.2018.03.014.

Grzegorczyk-Karolak I, Kuźma Ł, Wysokińska H. The effect of cytokinins on shoot proliferation, secondary metabolite production and antioxidant potential in shoot cultures of *Scutellaria alpina*. Plant Cell Tissue Organ Cult. 2015;122:699–708. doi: 10.1007/s11240-015-0804-5.

Guillon S, Tremouillaux-Guiller J, Pati PK, Rideau M, Ganetet P. Harnessing the potential of hairy roots: dawn of a new era. Trends Biotechnol. 2006;24:403–409. doi: 10.1016/j.tibtech.2006.07.002.

Gyulai G, Mester Z, Kiss J, Szemán L, Heszky L, Idnurm A. Somaclone breeding of reed canarygrass (*Phalaris arundinacea L*) Grass Forage Sci. 2003;58:210–215 doi: 10.1046/j.1365-2494.2003.00372.x.

Haberlandt G. Culturversuche mit isolierten PFlanzellen. Sitzungsber Akad Wiss Wien Math Nat. 1902;111:69–91.

Hayta S, Bayraktar M, Baykan erel S, Gurel A. Direct plant regeneration from different explants through micropropagation and determination of secondary metabolites in the critically endangered endemic *Rhaponticoides mykalea*. Plant Biosyst Int J Deal Asp Plant Biol. 2017;151:20–28. doi: 10.1080/11263504.2015.1057267

Ho TT, Lee KJ, Lee JD, Bhushan S, Paek KY, Park SY. Adventitious root culture of

*Polygonum multiflorum* for phenolic compounds and its pilot-scale production in 500 L-tank. Plant Cell Tissue Organ Cult. 2017;130:167–181. doi: 10.1007/s11240-017-1212-9.

Hu J, Gao X, Liu J, Xie C, Li J. Plant regeneration from petiole callus of *Amorphophallus albus* and analysis of somaclonal variation of regenerated plants by RAPD and ISSR markers. Bot Stud. 2008;49:189–197.

Huang B, Lin H, Yan C, Qiu H, Qiu L, Yu R. Optimal inductive and cultural conditions of *Polygonum multiflorum* transgenic hairy roots mediated with *Agrobacterium rhizogenes* R1601 and an analysis of their anthraquinone constituents. Pharmacogn Mag. 2014;10:77–82. doi: 10.4103/0973-1296.126671

Huang TK, McDonald KA. Bioreactor systems for in vitro production of foreign proteins using plant cell cultures. Biotechnol Adv. 2012;30:398–409. doi: 10.1016/j.biotechadv.2011.07.016.

Hussain S, Fareed S, Ansari S, Rahman A, Ahmad IZ, Saeed M. Current approaches toward production of secondary plant metabolites. J Pharm Bioallied Sci. 2012;4:10–20. doi: 10.4103/0975-7406.92725

Jamwal K, Bhattacharya S, Puri S. Plant growth regulator mediated consequences of secondary metabolites in medicinal plants. J Appl Res Med Aromat Plants. 2018 doi: 10.1016/j.jarmap.2017.12.003.

Jayarama Reddy (2014): Glossary of Modern Biotechnology: I.K.International Pvt Ltd. New Delhi.2014. ISBN: 9789382332756.

Jayarama Reddy (2008) Biotechnology of Orchids: 2008: I.K. International Pvt Ltd. New Delhi. ISBN:81-88237-67-1

Jayarama Reddy, Gnanasekaran D, Vijay D. and Ranganathan T.V.2010. In vitro studies on antiasthmatic, analgesic and anti-convulsant activities of the medicinal plant. *Bryonia laciniosa*. Linn. International Journal of Drug Discovery, ISSN: 0975–4423, Volume 2, Issue 2, 2010, pp-01-10.

Jayarama Reddy. (2016). *In Vitro* Organogenesis and Micropropagation of the Orchid Hybrid, *Cattleya* Naomi Kerns. European Journal of Biomedical and Pharmaceutical Sciences.Volume 3, Issue 9. Pp. 388-393

Jayarama Reddy. (2016). Micropropagation of *Dendrobium* Queen Sonia from Leaf Explants International Research Journal of Natural and Applied Sciences Vol.

3, Issue 8, August 2016 ISSN: 2349-4077. 7105 pp. 220-228

Jayarama Reddy. (2016). Nutrient Media Used For Micropropagation of Orchids: A Research Review. World Journal of Pharmaceutical Research Volume 5, Issue 9, pp. 1719-1732. DOI: 10.20959/wjpr20169-7036: ISSN 2277– 7105

Jayarama Reddy (2009). A Comprehensive Method to Isolate High Quality DNA from the Cultivars of Hibiscus; International Journal of Biotechnology Applications, Volume1, Issue 2, 2009. Pp-1-9.

Jayarama Reddy (2011. Protocol for the Economical Micropropagation of *Phalaenopsis* Queen )Emma Using Leaf Explants. International Journal of Biotechnology and Bioengineering Research (IJBBR)Volume 2 Number 2 (2011) pp. 195-206.

Juan-jie Z, Yue-sheng Y, Meng-fei L, Shu-qi L, Yi T, Han-bin C, Xiao-yang C. An efficient micropropagation protocol for direct organogenesis from leaf explants of an economically valuable plant, drumstick (*Moringa oleifera* Lam.) Ind Crops Prod. 2017;103:59–63. doi: 10.1016/j.indcrop.2017.03.028.

Karalija E, Ćavar Zeljković S, Tarkowski P, Muratović E, Parić A. The effect of cytokinins on growth, phenolics, antioxidant and antimicrobial potential in liquid agitated shoot cultures of *Knautia sarajevensis*. Plant Cell Tissue Organ Cult. 2017;131:347–357. doi: 10.1007/s11240-017-1288-2.

Karuppusamy S. A review on trends in production of secondary metabolites from higher plants by in vitro tissue, organ and cell cultures. J Med Plants Res. 2009;3:1222–1239.

Kaul S, Das S, Srivastava PS. Micropropagation of *Ajuga bracteosa*, a medicinal herb. Physiol Mol Biol Plants. 2013;19:289–296. doi: 10.1007/s12298-012-0161-3.

Klee H, Horsch R, Rogers S. Agrobacterium-mediated plant transformation and its further applications to plant biology. Annu rev Plant Physiol. 1987;38:467–486. doi: 10.1128/MMBR.67.1.16-37.2003.

Kodym A, Hollenthoner S, Zapata-Arias FJ. Cost reduction in the micropropagation of banana by using tubular skylights as source for natural lighting. In vitro Cell Dev Biol Plant. 2001;37:237–242. doi: 10.1007/s11627-001-0042-x.

Lubbe A, Verpoorte R. Cultivation of medicinal and aromatic plants for specialty industrial materials. Ind Crop Prod. 2011;34:785–801. doi: 10.1016/j.indcrop.2011.01.019.

Martínez-Estrada E, Caamal-Velázquez JH, Salina Ruíz J, Bello-Bello JJ. Assessment of somaclonal variation during sugarcane micropropagation in temporary inmersion bioreactors by intersimple sequence repeat (ISSR) markers. In vitro cell Dev Biol Plant. 2017;53:553–560. doi: 10.1007/s11627-017-9852-3.

Methora S, Goel MK, Kukreja AK, Mishra BN. Efficiency of liquid culture systems over conventional micropropagation: a progress towards commercialization. Afr J Biotechnol. 2007;6:1484–1492.

Nagella P, Murthy HN. Establishment of cell suspension cultures of *Withania somnifera* for the production of withanolide A. Bioresour Technol. 2010;101:6735–6739. doi: 10.1016/j.biortech.2010.03.078.

Pence VC. Evaluating costs for in vitro propagation and preservation of endangered plants. In Vitro Cell Dev Biol Plant. 2011;47:176–187. doi: 10.1007/s11627-010-9323-6.

Pniewski T, Czyz M, Wyrwa K, Bociag P, Krajewski P, Kapusta J. Micropropagation of transgenic lettuce containing HBsAg as a method of mass-scale production of standardised plant material for biofarming purposes. Plant Cell Rep. 2017;36:49–60. doi: 10.1007/s00299-016-2056-1.

Raduisene J, Karpaviciene B, Stanius Z. Effect of external and internal factors on secondary metabolites accumulation in St. John's wort. Bot Lith. 2012;18:101–108. doi: 10.2478/v10279-012-0012-8.

Rahman ZA, Noor ES, Ali MS, Mirad R, Othman AN. In vitro micropropagation of a valuable medicinal plant *Plectranthus amboinicus*. Am J Plant Sci. 2015;6:1091. doi: 10.4236/ajps.2015.68113.

Ramakrishna A, Ravishankar GA. Influence of abiotic stress signals on secondary metabolites in plants. Plant Signal Behav. 2011;6:1720–1731. doi: 10.4161/psb.6.11.17613

Rizvi MZ, Kukreja AK. In vitro propagation of an endangered medicinal herb *Chlorophytum borivilianum* Sant. et Fernand. through somatic embryogenesis. Physiol Mol Biol Plants. 2010;16:249–257. doi: 10.1007/s12298-010-0026-6.

Shekhawat MS, Kannan N, Manokari M. In vitro propagation of traditional medicinal and dye yielding plant *Morinda coreia* Buch.–Ham. S Afr J Bot. 2015;100:43–50. doi: 10.1016/j.sajb.2015.05.018.

Smetanska I. Production of secondary metabolites using plant cell cultures. Adv Biochem Eng Biotechnol. 2008;111:187–228. doi: 10.1007/10_2008_10.

Srivastava S, Srivastava AK. Hairy root culture for mass-production of high-value secondary metabolites. Crit Rev Biotechnol. 2007;27:29–43. doi: 10.1080/07388550601173918.

Thiruvengadam M, Praveen N, Kim EH, Kim SH, Chung IM. Production of anthraquinones, phenolic compounds and biological activities from hairy roots cultures of *Polygonum multiflorum* Thunb. Protoplasma. 2014;251:555–566. doi: 10.1007/s00709-013-0554-3.

Thorpe T. History of plant tissue culture. J Mol Microb Biotechnol. 2007;37:169–180.

Xu JF, Dolan MC, Medrano G, Cramer CL, Weathers PJ. Green factory: plants as bioproduction platforms for recombinant proteins. Biotechnol Adv. 2012;30:1171–1184. doi: 10.1016/j.biotechadv.2011.08.020.

Yamamoto Y. Tissue cultures of Euphorbia species. In: Hanover JW, Keathley DE, Wilson CM, Kuny G, editors. Genetic manipulation of woody plants. New York: Springer; 1988. pp. 365–376.

Yun UW, Yan Z, Amir R, Hong S, Jin YW, Lee EK, Loake GJ. Plant natural products: history, limitations and the potential of cambial meristematic cells. Biotechnol Genet Eng Rev. 2012;28:47–59. doi: 10.5661/bger-28-47.

Zuniga-Soto E, Mullins E, Didicova B. Ensifer-mediated transformation: an efficient non-Agrobacterium protocol for the genetic modification of rice. Springerplus. 2015;4:1–10. doi: 10.1186/s40064-015-1369-9.

Printed in the United States
by Baker & Taylor Publisher Services